Seminar Proceedings No. 25

WHEN THE GRASS IS GONE

Development Intervention in African Arid Lands

Edited by
P.T.W. Baxter

The Scandinavian Institute of African Studies, Uppsala 1991

Indexing terms
Development policy
Environmental effects
Environmental degradation
Pastoralism
Arid zone
Africa

Cover: Anna Bengtsson
Typesetting: Eva-Lena Volk
Copyediting: Sonja Johansson and Mai Palmberg

Printed in Sweden by
Bohusläningens Boktryckeri AB, Uddevalla 1991

ISSN 0281-0018
ISBN 91-7106-318-8

Contents

Foreword

Nordiska Afrikainstitutet (NAI), the Scandinavian Institute of African Studies, has an ongoing research programme under the general rubric "Human Life in African Arid Lands"; as part of its programme the institute has sponsored a number of seminars. This volume contains most of the papers presented at a seminar held at the Institute between 6 and 9 September, 1990. The participants were drawn from a range of disciplines; development studies, social anthropology, forestry, geography, administration, economics and veterinary studies; experiences ranged from the primarily academic to the primarily applied or administrative. There are relatively few opportunities for theorists and appliers actually to meet, let alone to exchange ideas and experiences. They are usually divided by a wall topped with razor wire, though it is not as high as the wall which divides developers from those being developed. But walls are there to be breached.

The Seminar was intense and exciting; in part because of the variety of experience sat around the table and, because most of the participants had been about long enough to know that there are a lot of questions but precious few answers. We were not brash enough to come to any positive conclusions; if answers to the questions could be so easily found then there would have been no need of a seminar. But, a number of common development errors and confusions were pointed out. However intractable the problems which development interventions provoke appeared to be, especially in fragile arid environments, everyone shared a determination to help limit environmental and human degradation. The first step is to help the people who live in the arid lands to manage their own futures. They know that their futures could be bright, because they know that their forbears managed very well in the drylands. Pastoralists have not been dubbed persistent for nothing. Given only half a chance they believe that they can wrest a bearable and independent life. The present populations have demonstrated their practical and cultural resilience just by surviving the disasters of the last two decades. Certainly few of us who write or read about arid zone pastoralists could have endured and survived as they have.

We were fortunate to have with us at the Seminar the following who enlarged and enriched our debates: — Ayele Gabre Merriam, Anders Hjort af Ornäs, Mohamed Ali Hussein, Leif Manger, Daniel Ndagala, Gisli Palsson, Michael Ståhl, Susanne Jacobsson and Johanna Palmberg.

Finally, all of us are grateful to Mohamed Salih for the kindly and efficient way in which he shepherded us all, and to the Nordiska Afrikainstitutet in Uppsala for sponsorship.

P.T.W. Baxter

Introduction

P.T.W. Baxter

> Know well the condition of your flocks, and
> give attention to your herds;
> for riches do not last forever;
> and does a crown endure to all generations?
> When the grass is gone, and the new growth appears,
> and the herbage of the mountains is gathered,
> the lambs will provide your clothing,
> and the goats the price of a field;
> there will be enough goats milk for your food,
> for the food of your household
> and maintenance of your maidens. (*Proverbs*, 27:23–27)

When Mohamed Salih invited me to participate in this seminar I accepted with alacrity, because the discussion of development policies and the environmental crisis were to be set in the context of their consequences for "Human Life"; that is their consequences for those Africans for whom the arid lands are their beloved homelands. We must all be distressed, to put it very mildly, about what has been happening, and is still happening, to the pastoral peoples of the world and perhaps particularly to those of northeastern Africa. The lack of concern that has been shown for their plight is extremely sad. Seminars and publications still mostly centre on rangeland problems, ecological problems, animal production problems, economic development problems—on almost every conceivable problem except any of those actually confronted by the people themselves. Indeed, the people themselves are often presented as the problem! Mohamed Salih has been one of the few scholars to stress that pastoral development programmes are quite different to programmes which assist pastoral people, and that pastoral development programmes actually often end up impoverishing and damaging their environment.

Most pastoral people are poorer, eat less well and have less control over their own lives than at any time since they recuperated from the cataclysmic run of droughts which ravaged East Africa and the Horn during the last two decades of the nineteenth century (Pankhurst and Johnson, 1988). When I first went to the Boran of northern Kenya in 1951, they were, as a people, infinitely better off in every material way than they are today. Even after the recent run of good rains, most of the population are not able to subsist by pastoral production, but still live in camps, in shanty towns, in the irrigation schemes or as depressed lumpenproletariat in the towns and cities. In parts of northern Kenya, and also in parts of Uganda, Ethiopia, Sudan and Somalia the situation has worsened progressively in recent years, because national governments have not been able (or not wished) to provide adequate security over wide stretches of rangeland. Consequently, these cannot be utilised because of the presence of armed bandits. Worse, in some countries the governments themselves have been the predators. Land and water have been alienated outright, or closed to pastoralists for use as irrigation projects (most of which have been costly flops), for dams, for towns and wild-life parks and, most shameful of all, for private ranches for the Wa Benzi, that new elite whose tribal mark is a limousine.

The grazing and the water available to the pastoral population has decreased as much by predation as by rain failures. As Johnson and Anderson point out, the "definition of ecological stress must include a political dimension" (1988:17), if not it is just a convenient lie (cf. Hjort, 1982). In summary, pastoralists have been increasingly marginalized, impoverished and oppressed over the last century and much of the rangeland has been degraded.

The essays in this volume, from the different viewpoints of their authors, are concerned with some of the consequences that development interventions have had for ordinary people, and with some of the environmental concomitants of particular interventions. There are some flickers of hope but mostly the essays provide hazard lights rather than guiding beacons. This is depressing, particularly when one recalls that development has been a major policy aim of the people themselves, of African governments and of international agencies for half a century. For example, the Colonial Development and Welfare Act was published by the United Kingdom government in 1940, during the Second World War. It had a budget of five million pounds sterling of which five per cent was allocated for research. Hopes of improving African economies and the lot of the African poor were high (Richards, 1977). Yet, half a century later, when colonial Africa is not even a memory for most Africans and hundreds of millions of pounds have been spent, we are still blundering.

Camilla Toulmin's opening essay is a succinct conspectus of planning processes for resource management in the countries of the Sahel: it examines the chasm which separates governmental perceptions of overriding national interests from the humble hopes and expectancies for simple survival held by ordinary pastoral families. A central problem is the overwhelming power, at least negatively and restrictively, that state governments have over the two crucial resources of land, i.e. pasture, and water: this is so because most of the arid and semi-arid zones, under various casuistical legal sleights, have been designated state assets. But, in practice, the states do not have the vision, nor the knowledge, nor the personnel to manage the natural resources over which they claim rights. They are trapped in their own grandiose rhetoric. Most state interventions are clumsy muddlings which produce neither revenue for the state nor income for the people. As far as I am aware not a single development intervention in pastoral areas has had any productive success at all. Yet, if the state did not have residual overseeing powers, the people would almost certainly have been devoured by local tyrants and predatory incomers. Governments, development agencies and the people themselves are confronted by dilemmas to which there are no clear answers. National land laws and policies frustrate resource planning at the local level because there are uncertainties as to where power over local resources actually resides in practice. So, we have an apparent paradox, progressive local planning for resource management may be least difficult where local "conservative" practices have survived ineffective interventions from the centre; that is where local resources are managed "traditionally" and there are local elders who retain some authority and can be empowered to take on fresh tasks. Toulmin's magisterial survey of past and current constraints to the development of local natural and human resources ends with a resumé of options for the future. She concludes, with guarded and cautious optimism that, given the national and political will and intelligent donor participation, local resource management might just be enabled to work effectively. If they do not the consequences do not bear imagining.

One theme that runs through most of the contributions is the need to take cognizance of indigenous knowledge and practices if local resource management and development plans are to work. The very fact that populations have increased, which is one reason for pressure on resources, is also evidence that indigenous practices are efficacious. Mohamed Salih analyses two series of development interventions in dairying: one was in Northern Nigeria and the other in Sudan. Both countries sought to satisfy the increasing demands of the urban elite for liquid milk by increasing national yields and, thereby, saving foreign currency by reducing imports of powdered and tinned milk. These aims, at first glance, seem sensible enough, but both government agencies sought to do this by importing technology and methods. The dairying practices and opinions of indigenous milk producers were entirely ignored. In both countries the development funds invested were entirely wasted, and in the Sudan the interventions actually did harm to the indigenous producers and seriously degraded the grazing areas within range of Khartoum. Fortunately, the Fulani dairywomen in Nigeria survived development interventions and continue to contribute to the feeding and comfort of the poorer urban population. Fulani dairywomen are able to sell processed milk products, to which their skilled labour gives added value, because development officials never took notice of their existence, so did not interfere with them. Mamman Aminu Ibrahim argues ebulliently for the relevance that indigenous agro-veterinary knowledge has for current problems, not only because it is often efficacious but because it is familiar and accessible. Further, innovations which are in accord with indigenous modes of perception and local practices are likely to be understood and utilized, whereas those based on unfamiliar, abstract, Western scientific practice are frequently ignored or, worse, misapplied. One alarming example, which was new to me, was the practice of spraying meat and fish with aerosol DDT sprays to kill off market flies! Technological innovation to be adopted successfully needs to be presented on a human scale and in terms that its recipients comprehend and can utilize. The plea to listen to, and to make use of, traditional local veterinary expertise is in line with current revaluations of indigenous agricultural and medical knowledge and with reassessments of indigenous experimentation. It is particularly relevant that the author has come to these conclusions by following his own route there. Mamman also demonstrates that African scientists are as likely as outside experts to ignore the evidence before their eyes because, so often, they wear the same blindfolds. A simple, but devastating, breakdown of the requests received for articles published by members of the staff of the Faculty of Veterinary Medicine at Ahmadu Bello University show that less than one per cent were from West Africa; this means that the very scholars who could reap most benefit from them, because they were working on similar problems and in similar environments, were the very people who ignored them. This suggests that the intellectual isolation about which African scholars rightly complain, is not only a product of currency shortages, which starve libraries and laboratories and limit academic exchanges, but also of the Eurocentric yearnings of many African scholars. The situation is made worse because it is often extremely difficult to get hold of African publications outside their country of origin, indeed it is often difficult even to get to know of their existence. African scholars then make insufficient use of African generated research and have limited access to European and American publications. This is obviously wasteful. Seminars such as this, in addition to bringing scholars together, also increase the transnational and transcontinental circulation of published knowledge.

The damaging consequences of outdated imported models are shown in George Keya's paper. This starts from the recognition which has been forced, at last, on the consciousness of development planners in Kenya, that imported inappropriate policies have failed the people and are themselves, in good part, responsible for the present environmental crisis. He surveys the old sad list of failed policies (grazing schemes, group ranches, commercial ranches and grazing blocks) and details the new approaches. The Kenyan National Development Plan which "appears to be a step in the right direction", makes proposals that are in line with current thinking on community participation, environmental protection, use of local expertise and appropriate technology. Though, as the author notes, the special funding required has still to be found, and it remains to be seen how the new approaches will work. One proposal does call for comment. The Plan suggests that pastoralists should seek additional sources of income to supplement that derived from livestock production, but the only positive suggestion is that they should collect and sell the aromatic gums which are exuded by certain desert trees. This is very odd, and suggests that high level planners have still not learned to converse with and learn from local people nor to use a library; the collection, trade and export of frankincense and other gums have gone on all over the Horn and across the Red Sea for centuries, and have been documented in several published studies. (Ahmed Yusuf Farah 1988, describes modes of collection, the trade etc. and discusses the literature. The gum is an important constituent of rituals among many peoples of the Horn.)

Odd Arnesen, a geographer, argues that in order to understand the arid zone crises we need to develop a theory of environmental degradation which includes the people, that is one which combines the insights of the natural sciences with those derived from sociology; this also requires the incorporation of the political organisation of space in relationship to ongoing social relationships, especially in so far as variations in the utilization of space depend on differential access to power. In the light of the theory that he adumbrates Arnesen looks at the restructuring of land systems, particularly in relation to collectivization and villageisation in Northern Ethiopia. Unfortunately, there is not much reliable data: interestingly it would seem that it is access to plough oxen which still makes the life of a peasant farmer just endurable.

Stig Johansson, a forester, examines the mostly avoidable environmental degradation which has accompanied large-scale irrigated agricultural projects in the Tana River Basin of Kenya. This, again, is a sad chronicle of technological arrogance and ignorant economic optimism riding roughshod over local people and their livelihoods. But, it is more than just a catalogue of errors and muddles, because it analyses why well meant endeavours by, presumably, well meaning experts went wrong, so it carries modest messages which should have damage limiting effects in the future. Michael Angstreich also writes of a forestry project but one, refreshingly, "that went right". It is a pity that the array of slides which accompanied his presentation cannot be reproduced here, because they are the evidence. He describes a cautiously conceived project which actively incorporated local people in an integrated range of water harvesting, well digging, dune stabilization and shelter belt tree planting in a denuded and degraded valley in Niger. The programme monitored its activities consistently and continuously, incorporated local people and appears to have been successful. It provides a useful and encouraging lesson in project design and management. Abu Sin's paper also concludes on an optimistic note. He argues that, after generations of either neglect or central authoritarian confusion in Central Butana, Sudan, a promising way forward has been found which promises (or did at the time of writing, how the present harrowing developments in the Sudan have affected developments we do not know) "a better quality of life for the people".

Mohamed Said Samatar and Jerome Gefu both discuss problems of pastoral production, the first in Somalia and the second in Nigeria. Samantar shows that the available figures for milk production in Somalia, a prerequisite for national planning in a country in which so many of the population are subsistence pastoralists, vary widely. So he endeavours to relate milk-production figures to population estimates. It would seem that the amount of milk available *per capita*, despite an increase in the camel population, is decreasing. The larger and tragic context of the bloody civil war which is raging must surely worsen this situation. Jerome Gefu argues that policies which have been based on the old thesis of the tragedy of the commons have no support from Nigerian experience. He concludes that developments which will ease the lot of the people and land regeneration both require management in which the people themselves actively participate. Finally, my own essay is an account of how Boran and Arssi, both Oromo speaking pastoralists, have organized the collection of salt-lick for their herds and flocks. It also suggests that developers should include human factors, such as sociability and the maintenance of neighbourhood solidarities, in their reckonings.

The consequences of the loss of stock and consequent destitution have generally been considered as if they were economic problems consequent on loss of livelihood. They are, of course, in part economic, but they are much more than that. A pastoralist's stock cannot simply

> ... be reduced to figures in a book-keeper's ledger, as if they were battery hens or pigs. Firstly, stock are components in a larger productive system that seeks above all to maintain the stock holding unit itself, that is, to maintain both the stock and the people who tend them and depend on them, and not just to turn a paper profit. Traditional modes of pastoral production have been modified to achieve this double task, of feeding and maintaining the family unit, most effectively. When rural development agencies ignore this social purpose and concentrate on the economic then their efforts founder. It was not chance that the peoples who have best survived the droughts of the last twenty years have generally been those who have been less 'developed', in the sense, that is, of being most oriented to cash production. Property is never inert, but where stock are incorporated into ongoing social life it assumes particular additional value, indeed it is transformed. Cattle have genealogies as well as pedigrees. A beast that has been allocated for bridewealth or sacrifice will have an ascertainable cash price, but that will be only part of its value. The contextual value of a beast is irrelevant to the butcher or to the rangeland manager. To measure pastoral production in 'standard livestock units' or in 'kilo calories' is convenient for some simple purposes, but generally it denatures, demeans, diminishes and distorts. It is the equivalent of categorising soldiers as 'replacements' or working people as 'hands', and can only be done by distancing the observer from life. Such detached observer categories legitimize exploitation by ignoring folk categories and concerns. As the Somali proverb puts it: 'To be without livestock is slavery.' (Baxter, 1990a, v)

To be stockless deprives a pastoralist of his or her identity and has endless cultural ramifications. The relationships between elders and juniors and between the genders (and especially between parents and children and between spouses) and between affines are quite altered.

The Boran, for example, in addition to economic loss have experienced bitter spiritual loss. They have not been able to maintain properly their full range of prayers and rituals, which require the regular sacrifice of stock, butter and coffee beans. Spiritual relationship have become as attenuated as social relationships. It is no wonder that the poor have become followers of Shaikh Nur Hussen of Bale. They have not only been deprived of sociability, conviviality and commensality, but of those realms of traditional life which were bound up with the rituals of sacrifice. They are left to a life of hardship and struggle bereft of the cheer and the sustenance of traditional rituals (Baxter,1987; 1990a and 1990b). But this is not an area of life with which development has ever been much concerned; unless it has been to proselytize. But, I would argue, development anthropology is not just a matter of understanding farming and productive systems and those social institutions which appear to be related to livestock production. The old truisms that man does not live by bread alone and that social anthropology is concerned with elucidating the interconnections between social institutions, may be hackneyed, but they are both still true. Equally trite but true is that the study of cultural change and persistence is where anthropologists should be best able to make a useful contribution, especially because of their knowledge of comparative ethnography, which enables them to set the particular in a generalizable context. Yet, as these essays demonstrate, their input to development programmes has been small. Why is that?

A good deal of excellent research had been carried out and published before the acute "crisis of pastoralism" developed in the seventies. In her admirable *Pastoral Nomads and Development*, published in 1975, Clare Oxby was able to cite a body of publications on Northeast Africa by the Dyson-Hudsons, Talal Asad, Ian Cunnison, Philip Gulliver, Ioan Lewis, Alan Jacobs, Paul Spencer and myself. Oxby was able to outline both the problems which confronted pastoralists and the policies which needed to be adopted, if pastoralists were to be given even half a chance. Enough data already existed to provide the base for sensible policies. But no one with managerial or decision-making influence has ever taken any cognizance of them. Pastoral development programmes then, as often they still do, started from scratch, as if nothing was known about the social systems or cultures of those about to be developed. Oxby argued for the following policies:

i. to control encroachment by farmers on land used by pastoral nomads;
ii. to protect both nomads' right of way through cultivated areas, and also farmers' crops;
iii. to draw up international agreements for the crossing of frontiers to the benefit of pastoral nomads in neighbouring countries;
iv. to ensure that pastoral nomads receive a just share of the revenue from the sale of animals for meat in relevant projects;

v. to phase out deep wells and boreholes where appropriate, in favour of a greater number of small traditional wells;
vi. to concentrate less on water-oriented projects, and more on pasture-oriented projects, for example, the storing of fodder;
vii. to set up mobile services, rather than encouraging nomads to come to the services of settled people: such as mobile schools and adult classes, mobile clinics and "bare-foot doctors", mobile veterinary services.

Oxby summarized: "The general message ... is that utmost caution is needed in the planning and implementation of development projects concerning pastoral nomads; in particular, one would hope that the decision-makers have learned all they can from the available literature on the subject, and the results of existing projects" (Oxby 1975: 8).

This summary is as pointed and relevant today as it was when she wrote it. That this is so, that so little seems to have changed is, at least at first glance, most dispiriting if not a reason for total despair.

Drought and famine and the very obvious collapse of most, probably all, pastoral development projects led to an efflorescence of anthropological studies of pastoralists, because simple economic or livestock production models had failed. This move was given an additional shove by the revived interest in social evolution among neo-Marxists (Baxter, 1990b). Everyone who participates in this symposium will be aware of that great upsurge in research, because they participated in it. Relief aid had come to be seen merely as a palliative and governmental and non-governmental agencies turned to development as a panacea. All the agencies recognized that they needed information, and many employed social anthropologists or funded research which had an anthropological component. Between the middle 1970s and 1980s there were many more anthropologists engaged in active research on the pastoral societies of north-eastern Africa, than there had been altogether over the previous seventy years; I think there have been at least seven PhD and several MA on Turkana alone, and heaven knows how many agency reports. A steady flow of good articles, symposia and monographs have been published regularly in Africa itself and the USA, UK, France, Germany, Holland, Italy, Norway,

Sweden and Canada. A critical bibliography, modelled on Clare Oxby's, is a pressing need to help bring us all up to date. Pastoral studies have flowered into one of the liveliest areas of development anthropology, and it may be that, just before it is too late, anthropological insights may yet influence development policies and practices. They have not done so very much to date, but there is at least a little reason to hope that they may yet.

Until very recently even when anthropologists have been employed on pastoral development programmes, their voices have not been much attended to. Indeed, though a good deal of lip service is now given to the contributions that anthropology could and should make, old preconceptions die hard. Now that so many pastoralists have become sedentary because they are destitute, it is difficult to sustain the argument (though it is still voiced) that sedentarization is a prerequisite for pastoral development: but other old canards about pastoralists still surface from time to time. In a very recent report from an agency with unimpeachable credentials, the following paragraph appears.

> ... livestock production is principally of a subsistence character. The pastoralists dispose of their stock unless needs for cash arise, such as school fees or buying grains and other food items during the dry months. Their cash requirements are small and their objective is not to produce surplus livestock for sale but to acquire wealth in the form of building up herds and flock size. However, there is an increasing pressure on these pastoralists to provide market stock, given the growing demand for meat in Kenya. (ANON, May 1990)

The accent is still on the production of beef for export to the cities, and pastoral production is seen only as a contribution to modernization. To feed oneself and to provide for oneself is not seen as a sufficient contribution to national development. This is, to me, an extraordinarily urban-centric point of view, as is the view that to build up a herd and flock rather than sell to the meat market is somehow backward. What would be said of a farmer who sold off his seed corn to buy luxuries? Why is it bad to live frugally? It reads like simple consumerism, though I am sure that is not the author's intention. In fact the people described have, for at least forty years, readily sold non-breeding stock to the market when they have been offered a fair price. Their herds do not contain high numbers of non-productive animals. The only true bit of the paragraph is that the pastoralists are under "increasing pressure ... to provide market stock". As so often the people are viewed as consumers and producers, not as people with particular values and needs. Nevertheless, I suggest that there are some grounds for restrained optimism. The diminishing effectiveness of most northeast African governments to protect their citizens, particularly those who live in the arid peripheries, also means that those governments may have less power to make damaging interventions. (*Survival International*, 1990 is the most recent publication to come to my notice which describes governmental predation; but almost daily reports from Somalia and Sudan confirm.)

Since the first days of colonial rule, most pastoralists have had to become past-masters at circumventing arbitrary interventions by governments. Government and close administration have never been anything other than an enemy. (This, of course, would also be the opinion of many rain-rich cultivators.) Weaker government could, in the long run, be less dangerous to pastoral survival than strong government is. Arid zone pastoralists also have generations of experience of surviving and recuperating from droughts (e.g. Sobania, 1988; 1990). I would bet that pastoralists will better survive another long drought, and a collapse of meat prices, than will the greedy WaBenzi who have alienated so much of the pastoralists' grazing and water for their own ranches and fattening grounds. (Though, I must admit, the data from Botswana does not encourage optimism; there the powerful have got much more powerful during the drought years—Gulbrandson, 1990 and Mazonde, 1990.) Western-style scientific management has become less sure of itself; traditional management practices are, at last, both in stock and crop production, seen as being ecologically friendly and long-term efficient. Bearing in mind the ominous predictions about the climatic consequences of the damage being done to the ozone layer Gowlett's words are worthy of repetition

> ... the developed countries are mainly in temperate regions, in which climatic changes take place on a different timescale from the tropics. This is a matter of chance, and it is quite likely that the developed countries would be no more prepared to cope with serious climatic change and its agricultural consequences than areas where it is operating now. (1988: 28)

Pastoralists are inured to tightening their belts. As long as enough groups of herders can survive and maintain widely ramifying networks of social relations which run across national, geographical and tribal boundaries, there are some grounds for hope (see Schlee, 1989). Cattle and small stock pastoralism has probably existed for five thousand years and cattle pastoralism for four thousand. Pastoralism has survived by forming, in various combinations, relationships with agriculture and with hunting and gathering. Pastoralists have demonstrated extraordinary endurance and adaptability, and deep management and political skills. Paradoxically, the very harshness of the environment has been a protection, because invaders and intruders have not had the skills to survive in it. I hear that, at present, an increasing number of Boran, Gabbra, Rendille and Turkana are re-establishing themselves in a reasonably prosperous pastoral life. Radical changes in economic and social stratification are occurring and the new pastoralists have become much more enmeshed, as buyers and sellers, in national and international markets than were their pre-drought predecessors (Baxter 1986, Hogg 1986). There are signs that the following prediction of John Gowlett may come true

> ... however great the Sahara is or may become, we can be sure that on the desert fringes there will be people herding their livestock and growing their crops. (1988: 45)

It has been clear to social anthropologists for a long time that indigenous knowledge, indigenous tenacity and indigenous social institutions have been, and are still, the keys to pastoral survival. It is encouraging to see, from several of these essays, that this is becoming an item in the new development orthodoxy (see Richards, 1985). Few, if any, anthropologists have ever argued for the romantic retention of outmoded social forms, although that is a common accusation made against them by Gradgrind-like development managers. No one could deny that improved water supplies and veterinary drugs have brought many benefits (though even these have been mixed blessings); but to go on from there to assume that methods of stock care and management, property and inheritance laws etc., should also be imported, is rank colonialist thinking as well as being profoundly unhistorical.

We really do have an excellent store of data and of interpretative studies of pastoral societies. Why does this gap between anthropological research findings and development planning persist? I suggest that it is because the people's perceptions of their needs, and the perceptions that governments and agencies have of the needs of the people not only do not coincide, but are mostly incompatible (e.g. Colson, 1982). Hence the constant opposition between, on the one hand, the people who are always disappointed by what government and agencies offer and, on the other hand, governments and agencies which are constantly indignant at the people's hostility and reluctant "conservatism". This is the great divide which often propels the social anthropologist into the role of advocate for the people, which then, in its turn, makes the anthropologist suspect to governments and agencies. The powerful, however liberal their words, are always suspicious of anthropologists. For example, in 1954 I applied to do fieldwork in British Somaliland. I was told that the then Governor, a liberal, dedicated and respected man, merely huffed that he would not have "a bloody anthropologist in the colony" because they were all disruptive, so I never got there!

The problem, then, is an old one and one which existed for the old Applied Anthropology, from which Development Anthropology is descended. It is apparent, I suggest, in the writings of C.G. Seligman, the principal author of *Pagan Tribes of the Nilotic Sudan* which was published in 1932, and provided the first summary of the information then available on the pastoral Nuer and Dinka. The research and the book were intended to be "the foundation of a scientific study of the peoples", throughout which the authors "constantly kept before us the necessities of the official" (Seligman and Seligman 1936: xi–xii). The Governor-General, Sir Harold MacMichael, who was himself a scholar and a strong supporter of anthropology, wrote a laudatory "Introduction" to the book, but there is no evidence at all that it was ever made use of by government as a tool for administration/development. Seligman was also the author of the entry "Applied Anthropology" in the 13th (1926) and 14th (1929) editions of the *Encyclopedia Britannica*. (I quote from the 14th edition because the 13th is not accessible to me. All the unattributed quotations below are from "Applied Anthropology".) Now Seligman was a scholar of what nowadays seems incredibly wide interests: he published on Africa, Ceylon, Melanesia, Java, China and the Ancient World and on many aspects of biological anthropology, archaeology, medicine, pathology, psychology and physiology. Applied Anthropology and the Sudan were only two of his many scientific interests, but for our present purposes they illustrate that the problems of what is now called development anthropology are not new.

At the time when Seligman was writing, pastoral areas in Africa were closed to development. In Kenya the northern pastoral areas were designated as closed or reserved areas. In the Sudan itself development centred on cotton and the Gezira Scheme. The pastoral Nuer and Dinka were still very actively resisting colonial rule. Applied Anthropology meant learning to understand native society in order that it could be administered with the minimum of expense and bloodshed. Seligman believed in science and in progress and that benevolent, paternalistic, colonial rule was the best way to bring both to backward subject peoples, be they Sudanese pastoralists or New Guinea headhunters. Many of Seligman's ideas and much of his vocabulary now seem anachronistic and offensively racist. He believed, for example, that race and psychology were interconnected; that white colonization was "a higher race imposing itself on the backward peoples", and that administrators should learn to "think Black". His "Hamitic theory" (really a speculation anyhow), which argued that "the Hamites were, in fact, the great civilizing force of black Africa" (Seligman 1930, (1939): 18) has long been a dead racist myth. (In passing, Seligman has been proven sadly wrong on another point. Defending the rights of subject peoples to enjoy their own rituals and pastimes without interference, he wrote: "What government at home would presume to stop ... public football matches?") But, in some ways, Seligman was not only ahead of his own times, but ahead of the present times.

He argued that officials and missionaries should have some anthropological training, so that when they intervened in native life they would be aware that social institutions were "interlocking". He stressed that: "Unforeseen and undesired ends have resulted from arbitrary interference with native custom." He stressed that much of what he termed "Positive Interference ...

undermines the native social structure of which administrators have remained entirely in ignorance." Unlike at least one development economist, he recognized that the administrator cannot "ignore magic and sorcery", and that "magic has its social as well as its anti-social side". He was aware of complexity.

But, as I wrote above, there is no evidence that he, nor even his distinguished protégé Evans-Pritchard, ever directly influenced government policies (Baxter, 1990b). And, perhaps, that was for the best, because although Seligman recognized "that the native has a point of view", which the researcher and administrator should seek to understand, there is little evidence that he thought that that point of view should influence policy designed to facilitate progress. Indeed, Evans Pritchard has argued that the Sudan government did not support anthropological research in the expectancy of any immediate practical return, but out of a sense of obligation to scholarship:

> The Sudan Government has ... for a long time and very generously supported anthropological research. In doing so it has allowed anthropologists to study pretty well where, what and how they liked. They have chosen the man and let him choose the plan. I think that they have been wise enough to do this because they have never been under the illusion that anything the anthropologist discovered was likely to have any great practical importance. They felt rather that a government ought to some extent to encourage scholarship, and they believed — and this is the point I want to stress — that a knowledge of the languages, cultures, and social life of the peoples of the Sudan has an immense value for administrative officials and others, quite apart from whether it solves any immediate practical problems or not. (Evans-Pritchard, 1951: 121)

Much has been written about "bottom-up" development, putting the last first and consultation but, in practice, the people have seldom got in even a whisper, let alone have their voices heard; whether expressed by themselves or as interpreted by others, the "point of view" of the people has not influenced development interventions. Indeed, many interventions have become confrontations, developers and those being developed not only speaking different languages, but having quite contradictory perceptions of what needs to be done. To bridge this gap, or chasm, takes diligence, patience, study and, above all, time and training. It is in his urging of the last two that Seligman was ahead of his time. He knew that there were no short cuts, and he also stressed that a "practical" field researcher required a grounding in theory. He wrote:

> Nothing suggests that the practical or applied outlook makes, or should make, any change in the method of the fieldworker. For if it be urged that attention to any particular theory is likely to distract the observer from a fair all-round study of facts, the reply is obvious that theory can only stimulate the fieldworker to fresh observations, which should neither limit his area of outlook nor discourage him from seeking the fullest knowledge of the folk he is investigating in their present and past environment. The more complete the apparatus of the observer ... the more likely is he to form a complete picture of the people .... .

Now long field research, careful scientific procedures and, perhaps above all, theory, seem to be anathema to practical development experts. Even those who see merit in the findings of social anthropology usually claim that the urgency of the job in hand precludes theory and long research and considered reporting. This, I suggest, is why so little use has been made of anthropology by developers. They are just in too much of a hurry to take cognizance of the complexities of anthropological findings, and they have just not had sufficient anthropological induction themselves to understand or to give value to those findings. They may no longer believe in the quick technological fix, but they still seek the quickly implementable idea and the simpler the better.

To conclude I will quote from two eminent authorities whose work I admire, and who both endeavour to incorporate anthropological findings in their work. If they miss the point, then those who do not try will never see there is a point at all.

Firstly, Robert Chambers writes in *Rural Development: Putting the Last First*:

> ... social anthropologists who have spent any length of time in the field usually know a great deal and have many insights useful to practitioners and interesting to academics. But often they do not know what they know, and their self-isolation hinders these exchanges .... (1983: 61)

Self-isolation indeed! I have only ever been asked to provide brief and simple answers to complex questions. It is as if they had a Winston Churchill complex and wanted every issue presented on one side of one sheet of paper. Development workers should not be snappers-up of the unconsidered trifles which anthropologists collect. Social anthropology is not a hedgerow from which facts can be plucked. It is a discipline, a way of looking at facts. Secondly, Stephen Sandford's already classic magisterial survey *Management of Pastoral Development in the Third World* relegates "organization of pastoralists" to Chapter 10, and then, as in the earlier chapters, pastoralists are considered almost as an undifferentiated social category. The great collection of anthropological analyses are only rifled for trifles. If Chambers and Sandford, who are both so dedicated to the improvement of the lot of the poor, cannot make the effort to understand an anthropological monograph, then maybe my optimism is unjustified, and we shall never bridge the gap. But, on the other hand, certainly anthropological insights have penetrated parts of USAID, ODA and most conspicuously all the branches, governmental and non-governmental, of Scandinavian development programmes. So we must keep on trying. If we sometimes seem to be repeating old messages, it is only because those for whom they were meant have refused to listen. But, with or without us, given a bit of luck, the pastoralists will manage anyhow. Certainly, we still have more to learn from them than they do from us.

REFERENCES

Ahmed Yusuf Farah, 1988: *The Milk of the Boswellia Forests: Frankincense Production among Pastoral Somali*, PhD Thesis, University of London.

Baxter P.T.W., 1987: "Some Observations on the Short Hymns Sung in Praise of Shaikh Nur Hussein of Bale", in Ahmed el Shahi (ed.), *The Diversity of the Muslim Community*, Ithaca Press, 139–152.

Baxter P.T.W., 1988: "The 'New' East African Pastoralism: An Overview", Edinburgh Anthropology , 2 (*Munro Lectures* 1985).

Baxter P.T.W., 1990a: "Introduction" to Baxter and Hogg (eds.), i–vii.

Baxter P.T.W.,1990b: "Oromo Blessings and Greetings", in Anita Jacobson-Widding and Walter van Beck (eds), *The Creative Communion: African Folk Models of Fertility and the Regeneration of Life*. Uppsala Studies in Cultural Anthropology, Almqvist and Wiksell, Stockholm.

Baxter P.T.W., 1990a: "Social Anthropology and Pastoral Development in North East Africa", in *Ethnography and Current Questions of our Time*. A symposium held on February 27 & 28 1990 at Karl Marx University, Leipzig.

Baxter P.T.W. with Hogg, R. (eds.), 1990: *Property, Poverty and People: Changing Rights in Property and the Problem of Pastoral Development*, The Department of Social Anthropology and the International Development Centre, University of Manchester.

Chambers Robert, 1983: *Rural Development: Putting the Last First*, London, Longmans.

Colson Elizabeth, 1982: *Planned Change: The Creation of a New Community*. The 27th Bernard Moses Memorial Lecture, The Institute of International Studies, Berkeley.

Evans-Pritchard E.E., 1951: *Social Anthropology*, London: Cohen and West.

Gowlett John A.J., 1988: "Human Adapation and Long-term Climatic Change in North-East Africa: An Archaeological Perspective", in Douglas Johnson and David Anderson (eds.), 27–45.

Gulbrandson Ornulf, 1990: "Material and Symbolic Aspects of Changing Property Relations in a Tswana Society", in Baxter and Hogg (eds.), 164–181.

Hjort Anders, 1982: "A Critique of Ecological Models of Pastoral Land Use", *Nomadic Peoples*, 10, 11–21.

Hogg Richard, 1986: "The New Pastoralism: Poverty and Dependency in Northern Kenya", *Africa*, 56(3), 319–333.

Johnson Douglas and David Anderson (eds.), 1988: *The Ecology of Survival: Case Studies from Northeast African History*, London, Lester Crook Academic Publishing.

Little P. D. and M. M. Horowitz (eds), 1987: *Lands at Risk in the Third World: Local Level Perspectives*. IDA Monographs in Development Anthropology. Boulder and London, Westview Press.

Mazonde Isaac, 1990: "From Communal Water Points to Private Wells and Boreholes in Botswana's Communal Areas", in Baxter and Hogg (eds.), 182–191.

Oxby Clare 1975: *Pastoral Nomads and Development*, London, The International African Institute.

Pankhurst Richard and Douglas Johnson, 1988: "The Great Drought and Famine of 1888–92 in Northeast Africa", in D. Johnson and D. Anderson (eds.) (1988).

Richards Paul, 1985: *Indigenous Agricultural Revolution*, London, Hutchinson.

Richards Audry, I., 1977: "The Colonial Office and the Organization of Social Research", *Anthropological Forum*, VI (2):33–53.

Sandford S., 1983: *Management of Pastoral Development in the Third World*. Chichester ODI and John Wiley and Sons.

Schlee Gunther, 1989: *Identities on the Move: Clanship and Pastoralism in Northern Kenya*, Manchester University Press for the International African Institute.

Seligman C.G., 1913: "Some Aspects of the Hamitic Problem in the Anglo-Egyptian Sudan", *J.R.A.I.*, 43, 593–705.

Seligman C.G., 1930 (1939): *Races of Africa*, London: Thornton Butterworth.

Seligman C.G., 1929: "Applied Anthropology", Encyclopedia Britannica.

Seligman C.G. and Seligman, B.Z., 1932: *Pagan Tribes of the Nilotic Sudan*, London, Routledge.

Sobania Neal, 1990: "Pastoralist Migration and Colonial Policy: A Case Study from Northern Kenya", in Douglas Johnson and David Anderson, 219–240.

Sobania Neal, 1990: "Social Relationships as an Aspect of Property Rights: North Kenya in the Pre-colonial and Colonial Periods", in P.T.W. Baxter and Richard Hogg (eds), 1990, 1–19.

Survival International, 1990: *Unquiet Pastures: The Nomadic Peoples of North East Kenya Today*.

Bridging the Gap Between Top-Down and Bottom-Up in Natural Resource Management

Camilla Toulmin

This paper examines the different levels at which natural resource management decisions are taken in the Sahel, from national plans to combat desertification to village committees for allocating farmland. It looks at the variety of tenure systems currently operating, the contradictions between these and the general difficulties faced in bridging the gap between local and national interests and capacities in the management of land, water trees and grazing. It concludes with an assessment of current changes in resource management systems and the conditions for their effectiveness.

NATIONAL PLANNING

Natural resource planning at the national level is usually made up of several components:

Government ministries concerned with agriculture, livestock, water and environment, departments of water, forestry and hunting, all operate in this field in a variety of configurations, supplemented by units within the ministry of justice dealing with land law and mapping departments concerned with surveys. Ministries of wildlife and tourism may also have important roles to play in certain regions where their interests are concerned. Environment ministries usually have limited strength within this net of interlocking interests and conflicting responsibilities.

Most countries have a welter of national plans and programmes for action in the natural resource field, such as National Plans of Action to Combat Desertification, National Conservation Strategies, Tropical Forestry Action Plans, in addition to National Development plans, energy plans, and so on. The demand and funding for these programming exercises frequently come from outside donors, each donor having an interest in a particular kind of planning exercise and document. Overall, the main impression is one of limited coordination amongst donors in terms of the kind of support given and the programming exercises undertaken. Competition amongst donors leads to the duplication of work and the absorption of much host government time in consultation with visiting missions of experts. For example, it was reckoned that Burkina Faso received more than 700 expert missions in 1988, each one tying-up government staff and resources.

The bias in much national resource planning is towards physical planning, mapping, assessment of carrying capacities and so on rather than on institutional feasibility, so that physical targets are set without reference to the human dimension. Beautiful, multicoloured land use assessment surveys are carried out to demonstrate the suitability of different land areas for different uses. It is unclear how this material can be made use of at ground-level, since planners must cope with people already using land in ways which may not be consistent with the mapping exercise.

At regional and local level, a similar pattern of activity is found, with the bilateral funder's role taken by the large number of non-governmental organisations (NGOs) now involved in environmental and natural resource management projects. Similar problems of competition and overlap arise.

TENURE SYSTEMS VESTED IN THE STATE

It is at the national level that formal tenure over the country's natural resources has been held since colonial times. This tenure system stems from colonial legislation and has been subsequently modified following independence. In general these systems have involved the nationalization of land and water resources, while allocating rights of usufruct to traditional users. These rights usually refer to land currently under cultivation; by contrast, right of farmers and herders to land in fallow and to traditional grazing territory are far less clearly defined and in practice this has permitted in many cases the alienation of these resources from customary users. Formal ownership of the land by the state allows the authorities to gain access to particular pieces of land when necessary, for example to carry out an irrigation or mining project.

Parallel to state ownership of the land is the state's right and duty to manage and plan how resources should be used. In practice, national and local administrators have rarely been able to carry out these functions effectively. The very large areas of land involved, their scattered populations, and the weak capacity of local administrations have combined to create a power vacuum on the ground. Local communities have lost the legal power to decide and enforce decisions about rights of access to local resources, yet this loss has not been replaced by a state provided system at local level able to ensure firm and fair allocation of rights when brought into dispute.

TENURE SYSTEMS VESTED IN THE COMMUNITY

The second system of tenure which exists in some form through much of the Sahel is based on the traditional pattern for managing resources found in much of Africa, whereby access to land is obtained through membership of a given community which has rights of custody over a given area. Generally such land rights are acknowledged to derive from settlement, so that the first families to establish a farming hamlet are the ones to exercise subsequent control over who may gain access to land. Traditional rules of tenure ensure rights of usufruct to land for so long as the household maintains its other obligations to the community by, for example, making the necessary religious sacrifices at the start of the farming season. Rights to resources other than land (such as water from communal village wells) may be contingent on membership of the community, while others (such as hunting) may require special authorisation by the appropriate body (such as the local hunters' association). Ownership is more tightly defined where a family has invested time and effort in raising the productivity of a particular resource; for example, families may claim ownership of certain trees in the village's woodland when they have been pruning these trees regularly, while ownership of a well too is usually attributed to the person or group responsible for having dug it.

These traditional tenure systems stem from pre-colonial times; nevertheless, the operation and acceptance was not always free of conflict both internally and externally. Higher authorities could be called in to defend rights where these were not adhered to by others. Only within the central heartlands of precolonial states were rights likely to have been firmly enforced, while on the edges there was much insecurity surrounding which communities had rights to what leading to raids between opposing sides.

TENURE IN PRACTICE

Despite the state's formal right to manage all resources, in practice through much of the Sahel local communities continue to exercise some of the functions they held under traditional tenure. Local administrators, recognizing the extreme weakness of their own resources, may leave communities to manage and police their own affairs. Communities themselves may do their best to avoid involvement with the local administration, even where there is a serious conflict, since to call in government is likely to lead to very heavy costs, in terms of payments to the appropriate officials and the ultimate fine which may be imposed on both sides to the dispute. However, in certain areas and increasingly throughout the region, all resources are coming under mounting pressure, particularly higher potential farming and grazing land, trees and pastures in more accessible areas close to urban markets, etc. In these cases, where the stakes may be high and the returns to gaining secure rights of access great, the state is likely to intervene, to avoid conflict and to allocate rights according to certain principles. Officially, these principles usually refer to the ability of the person demanding the resource being able to put the land to good use ("la mise en valeur"), giving priority to those able to demonstrate access to labour, capital, and other resources. "Making good use of land" tends to be interpreted to favour farming, rather than grazing. In this way, throughout dryland Africa, herders have lost access to prime grazing lands in low-lying areas (*fadama, bas-fonds, wadis*) to make room for dry season gardens, rice and wheat cultivation. Principles of allocation also benefit those able to manipulate and make use of contacts within the administration, to acquire rights in priority over other potential users.

The main factors responsible for this increased pressure on resources, and the increasing disorder associated with their management and use, are common to most other areas of dryland in Africa and include:

- the prolonged period of drought over the past 25 years
- population growth and the expansion of cultivated areas, aided by the introduction of animal traction and mechanised farming techniques
- rising levels of conflict and insecurity which have forced people to concentrate in certain safer areas, where degradation intensifies while insecure areas are under-used
- the lack of clarity regarding by whom and how decisions about land and water use are taken, due to the breakdown in traditional authority over the allocation of use rights and government's incapacity to carry out these functions effectively
- rising demand for certain commodities and their increased commercial value, such as charcoal, fuelwood and fodder.

ADVERSE EFFECTS OF CURRENT TENURE SYSTEMS ON LAND HUSBANDRY

The state's right to control and allocate land which is not currently under cultivation takes from local communities the incentive to manage and set aside land for sylvipastoral use and for future farm land. There is an increasing trend for farmers to develop highly extensive forms of land use, to demarcate for themselves a large area which they can demonstrate as being under cultivation, however rudimentary it be. In a similar fashion, herders have felt it necessary to start cultivating land, in however vestigial a fashion, in order to demonstrate to the authorities that they have certain rights to a piece of land, since customary tenure to grazing land has rarely been considered to carry the same weight as that of farmland. These forms of land use are often highly damaging to the vegetation and soils as land may be cleared of trees and the soil exposed but little viable crop cover developed, which increases risks of soil erosion.

The state also retains the right to appropriate and re-allocate cultivated land in some cases. The importance of this right has received considerable attention in the development literature, and in certain contexts it may prove a powerful disincentive to good land management. For example, on irrigation schemes, farmers usually have very weak rights of tenure which strongly discourage their investment of labour in improving the land. However, other writers have maintained that in practice farmers have rarely been discouraged from making investments in manuring and in building soil conservation works by this lack of firm title, as they are confident that their rights to their farm land would be supported in the event of any conflict.

Tree tenure presents particular problems. In most countries, trees can only be cut down on possession of a permit from the relevant authority, even when it "belongs" to a particular person who either planted it or on whose land it is growing. This permit must be obtained from the forestry officer at the local government headquarters and usually costs a significant sum. Even those who have planted a tree must obtain official licence to harvest it, in theory. In practice, of course, such legislation is not enforceable, but it is probably sufficiently discouraging to prevent some tree planting from taking place. The existence of this process and the accompanying legislation also provides an endless source of conflict between forest guards and local people, the former able to impose substantial fines on the latter where there is suspicion of their being involved in woodcutting.

Pastoral land tenure has been the subject of great debate over many decades, with competing models for managing these extensive and highly variable resources. Solutions proposed in the past have ranged from privatization through the development of ranches to attempts to allocate rights to a certain group of herders. Currently, tenure systems betray a mixture of open access regimes, particularly around public bore-holes, and more demarcated regimes where certain herders have been able to maintain some rights to control access to water and hence to surrounding pasture lands. In areas of particular value, such as the inner Niger Delta in Mali, remnants of the traditional pre-colonial system for controlling access to flood plain pastures have been used to establish a new model. However here, as in other areas of high value, it is clear that establishment of contacts within the administration of this tenure system may confer unfair advantages in helping to guarantee prior access by certain herds to these high quality grazing lands.

The pastoral sector throughout the Sahel has been badly hit by successive periods of drought, the continued rapid spread of farmers into former grazing areas, impoverishment of herding families by the loss of their livestock capital and poor terms of trade, and the progressive involvement within the livestock sector of outside investors, leading to the loss of control over herds and pastures. Herders have not been able to defend their customary rights to grazing lands against the incursions of farmers; in cases of conflict over land, judgement is almost always made in favour of cultivators. The formal ownership and management of water and pasture by the state have led to even greater problems of controlling access and use of these resources than in farming areas, due to the vast areas involved and the practical impossibility of the government "policing" access to water and grazing.

Cynics view the current situation regarding land rights, and their lack of clarity as a conscious strategy by national and local decision-makers. Governments use the current anarchy in land management and use to justify continued central control, on the grounds that local populations are evidently incapable of managing their own resources. Local administrators help create further chaos by arbitrary decisions regarding who has access to certain resources, and thus ensure that contesting parties will hope to ensure a decision is made in their favour by providing large bribes. Clear land legislation would damage the interests of these administrators by removing from them the power to allocate as they see fit. As far as pastoral tenure is concerned, any devolution of power over water and grazing to politically marginal pastoral groups is seen as very unlikely, as this would strengthen their autonomy, seen as a threat to central government power.

POLICY CHANGES: RHETORIC OR REAL?

Since the early 1980s, governments and donors have talked increasingly of shifting to local communities the rights to manage resources within their territory. Government documents have laid out policies leading to the "responsabilization" of farming and herding populations and all involved have peppered their strategy and project documents with commitments to peoples' participation in the design of projects and in managing their resources. It all seems very encouraging. Work is proceeding on the definition of land tenure codes in several countries which, it is hoped, will provide the legal basis for communities to plan and manage how they use their land, with the guarantee, in theory, that the state will enforce their rights should they be contested. Several countries have begun pilot schemes for the management of village lands. However, there are still many questions about this process of shifting responsibility and power to local communities which need to be considered. It must also be recognized that devolution of control to the local level is no guarantee of fair access rules and wise decision-making, and begs a number of questions, which include the following:

i. At what level is the local community to be defined — the village, or group of villages, the herding camp, or herders' associations? Must new structures be set up or can control be vested in existing social organisations? Given the heterogeneous nature of all communities, is there a representative body at the local level able to manage and control access to resources fairly?
ii. If power to manage resources is vested in each village or herders' group, what support will they need from local government to provide, first, the rapid and effective enforcement of their rights when these are brought into question and, second, the technical support and skills required to make the most of these resources? Can local government provide this support? Often they will not be able to do so.

iii. The interests of local communities and national government may not always coincide. Examples include land of great economic or strategic interest to the state, or of outstanding natural or scientific value. Whose interests should receive priority and how should this be decided? Management structures need to be evolved which bring together local representatives and national government to agree to management rules in cases of this sort. A particular case where local and national interests are at odds concerns the settlement of migrant farmers moving from one part of the country to another in search of better land. For example, the cotton-growing region of Burkina Faso has seen a huge in-migration of farmers from drier zones further north. As a result, many villagers have seen their reserves of fallow and grazing land swallowed up. What rights should indigenous populations have to set the terms on which outsiders can obtain access to their land, water, trees and pasture? And, if indigenous populations do not have firm control over rights of access, what incentive do they have for investing in the careful management of these resources?

A pessimistic view of the trend towards "responsabilization" of local communities might see various forces at work, such as the following. Governments have been pushed by donors to talk about greater participation by their populations in the development process and in the management of their own resources. The "new conditionality" and pressure for multiparty democracy now coming from many bilateral funders are further recent trends in this direction. However, governments have been skilled in handling the development discourse and in emphasizing the terms and approaches currently in vogue amongst donors. In reality, governments have made little if any progress in transferring real power away from the centre. The only "responsabilization" that will take place, according to this view, is likely to concern cases where the state can divest itself from certain costly obligations, for example by transferring responsibility for maintenance of bore-holes to pastoral associations, and by handing over the role of maintaining irrigation schemes to water user groups. "Responsibility" for funding maintenance costs is unlikely to be rewarded by the transfer of real decision-making powers where these run counter to current or future interests of the government.

A more generous interpretation of current policy in much of the Sahel sees a growing number of government officials as increasingly aware of the need to find solutions to land tenure issues and to develop a closer dialogue with their populations. Transferring rights and responsibilities back to local users may be the beginnings of a solution. However, it is a solution which still requires an important role for local and national government in the guaranteeing and enforcement of rights, in the technical support required and in the balancing of local versus outsider and national interests. The experience of the next few years should establish whether there is the political will necessary at national level and the resources available from the donor community to enable such changes in resource management by local users to work effectively.

Livestock Development or Pastoral Development?

M.A. Mohamed Salih

INTRODUCTION

This paper deals with pastoral development related policies in the Sudan and Nigeria with the prime interest of distinguishing between notions such as livestock development and pastoral development and their implications for pastoral communities. It also attempts to assess the effect of the introduction of modern dairy farms on small-scale milk producers, and the patterns and trends of recent pastoral production systems in the arid and semi arid lands of the Sudan and Nigeria. The basic assumption here is that planners, administrators and many social scientists still hang on to the belief that pastoral development and livestock development policies are synonymous. However, it is argued here that it is time to recognize that there are several differences between these two concepts which imply different attitudes and perceptions. Whereas livestock development is delineated as an economic activity governed by the rules of the market economy, pastoral development is a social activity geared towards "developing" the pastoralists. Sharp as it may be, the border line between these two notions is very important since policies to develop livestock are not always identical with pastoral development and *vice versa*.

Livestock development is more specific and more clearly defined than pastoral development. According to contemporary literature (Raikes, 1981, Sandford, 1983) policies intended for livestock development often include the following:

1. Technical advancement and the introduction of new inputs, medicine and vaccine for disease control, engine power for water drilling, farming of fodder, etc.

2. Specialization and diversification of livestock products through the adoption of the so-called superior production and management techniques. There is also a tendency towards specialization in livestock products, for example, meat, dairy, wool, hide, etc.
3. Increasing the capacity to evade seasonality by transforming perishable products such as milk into storable forms often aided by technical advancement and specialization.
4. Developing an integrated marketing outlet responsive to the demands of national and international consumers. The development of livestock marketing boards and the emergence of middlemen and wholesale traders operating at the local, regional and national markets.

The overall objective of livestock industry is to increase livestock production and productivity, maintain steady off-take and provide low priced livestock products to satisfy the needs of the more politically vocal urban dwellers. In other words, pastoralists are seen as mere keepers of livestock, providers of cheap livestock products and an indispensable source of revenue to the national treasury.

On the other hand, pastoral development policies are loosely defined but, ideally, they have more to do with introducing small-scale projects and an input delivery system specifically suited to the needs of pastoral societies. The prime objectives of such policies are to enhance social development and improve living standards. Pastoral development policies, therefore, are best understood as an attempt to facilitate development among pastoral societies with due recognition to the importance of integrating aspects of modern knowledge to the traditional system of production. Livestock plans, on the other hand, have much in common with pastoral development even though they are not directly targeted to benefit pastoralists, except in the case of the provision of water and health facilities. Livestock development policies presume that a trickle-down effect will eventually diffuse economic benefits and raise the living standards of the pastoralists once such policies are implemented. In essence, this postulate encompasses two fallacies: first, that pastoral societies would accept this strategy and second, an increase in livestock would ultimately result in an increased off-take. The two assumptions have proved to be wrong; many pastoral societies, not without good reason, were suspicious of the programmes and at times refrained from being party to them and second, few sales were made to fulfil the market demand for more livestock and livestock products. Administrators and planners associate pastoral development with livestock development, and one of their earliest policy objectives was that settlement is a prerequisite to pastoral development. This paper is devoted to the discussion of these issues in relation to the Sudan and Nigeria and the impact of livestock development policies on pastoral societies.

LIVESTOCK DEVELOPMENT AND POLICIES IN PRACTICE

Pastoralists represent 14.1 per cent of the total population of 20.6 million Sudanese and contribute over 12 per cent of the GDP and 23.5 per cent of foreign exchange earnings from the export of live and slaughtered animals and livestock products, such as hides. In Nigeria the Fulani, who constitute the main producers of livestock and livestock products were about 4.8 million in 1965 and represent about 8.6 per cent of the total population. According to the Green Revolution (1983) pastoralists contribute less than 5 per cent of the GDP or 27.4 per cent of the total share of the agricultural sector.

As in the case of Nigeria, the notion of livestock development was not fully formulated in the Sudan until the 1970s, when an increasing demand for fresh milk and meat for an expanding urban population, on the one hand, and pastoralists' need for veterinary services, disease control and vaccination campaigns, on the other, meant that some sort of policy had to be formulated. The Sudan's Ten Year Plan (1961–1970) was biased to the development of the agricultural and industrial sectors, but it showed some interest in livestock development. This period witnessed a considerable increase in veterinary services and some livestock development projects.

Three livestock development centres were established to improve Kenana and Butana dairy cattle breeds, to develop mixed farming, assist nomads to appreciate better methods of livestock raising, to distribute males of improved breeds and to advise the nomads on better and more profitable livestock raising practices.

This was followed by the establishment of five centres for the improvement of dairy production in large towns (El-Obeid, Juba, Malakal, Torit, Atbara). Other important developments during this period were in the field of veterinary medicine, the production of vaccines and the establishment of 130 animal dispensaries and 20 veterinary hospitals throughout the country. Towards the end of the plan in 1970 higher fertility rates were recorded as a direct result of the introduction of modern medicine, vaccination and disease control campaigns. These centres did not benefit pastoralists and remained isolated from the bulk of the small producers and had no real connection to the wealthy owners.

The Five Year Plan (1971–1975) extended these activities and established seven major settlement areas for mixed farming and ranching. These settlement areas also provided watering points and corridors for those animals migrating mainly from Dar Missiriya District, Kordofan Province to encourage the Baggara to settle around Babanusa Milk Factory. However, the 1972–75 drought forced the Baggara to change their migratory pattern, and few households were able to endure settlement while their herds were declining rapidly due to drought and shortage of pasture around the boreholes. The factory's capacity was underutilized, there was a shortage of spare parts, and it suffered huge economic losses which rendered it useless by 1978.

Towards the end of the plan, in 1974/75, the Animal Production Corporation was established in order to promote the development of private and public dairy, fattening and poultry projects. These were to be located around the large towns with the main objective of providing cheap dairy products to urban dwellers. According to this policy ten government dairy farms were established in the principal towns to supply fresh milk, and to advise prospective farmers on how to run dairy farms on a profitable basis. This experiment was extended to the Gezira scheme in 1973, as part of the Government policy to integrate livestock and agricultural production. About 39 dairy farms and fattening units were established in Gezira in 1976, benefiting from the Nesheishiba livestock improvement centre which provided and administered artificial insemination to reproduce improved Kenana breeds.

The steady rise of meat prices in the capital towns (Khartoum, Omdurman and Khartoum North) prompted the development of rest-stations throughout the length of the livestock route from Western Sudan to Omdurman town. An Animal Production Corporation was established and 46 rest-stations were developed to provide water, fodder and limited veterinary services. The project was also connected with a plan to establish a disease-free zone guarded by veterinary police. Holding grounds were also established around Omdurman and other towns to increase livestock supply both for local use and for export. The main objective of the project was to overcome the problem of livestock moving in search of water and fodder away from areas of demand during the dry season. The big livestock merchants and traders used the holding grounds to hold their livestock back from the market until the prices soared and only sold small numbers. It also offered the livestock traders operating in the export sector the opportunity to diversify their outlets and compete favourably with the local slaughter-houses. This prompted a decision by the Government in 1975/76 to impose a ban on animal exports in order to satisfy the demands of the local market. With the open Sudanese borders and the availability of water transport across the Red Sea, smuggling was intensified and the Government lost a valuable source of revenue.

The Government's attempt to offset domestic and export demands met a new challenge in 1985/86 after the drought. It appeared that pastoralists withheld animal sales in order to build up their herds. Grain prices declined and fewer animals were sold to satisfy household needs. Although there was no declared policy on the issue, various statements by Sudanese officials offered three interrelated options: first, to continue with the banning of livestock exports specially females; second, to introduce pricing measures and curb the black market; and third, to eliminate middlemen and improve producer prices. The first option was neither practical nor desired, since livestock proved to be the only profitable sector of the economy as far as the cost of production was concerned. The Government reaps about 80 million dollars annually in foreign exchange earnings from a very meagre investment in the livestock sector. Second, the economic recession was hitting hard, while the agricultural sector was in shambles consuming 65 per cent of the cost of production in the importation of chemicals and fertilizers, which rendered it non-profitable.

The price controls failed to realize their objectives and, for the first time in Sudanese history, butchers waged several strikes against pricing controls. Their point of view was that livestock traders and middlemen made over 50 per cent profit from the sale of every livestock unit and that they were the real cause of the highly priced meat. The 1983 ordinance, which banned the sale of meat on Wednesdays, was reinforced to slow down the demand for meat. However, people began to store larger quantities of meat to compensate for that, and the demand remained unchanged.

The efforts to develop modern dairy production in Nigeria were described by Water-Bayers in 1988. This study reveals that Nigeria was relegated from an exporter of dairy products during the 1940s to a net importer from the 1960s onward. From the mid 1970s, the Third National Development Plan established ten factories for reconstituting imported powdered milk. The attempt to establish small dairy farms among the Fulani pastoralists during the same period failed because of what Waters-Bayer (1988:35) describes as

> ... disagreement among planners on the aim of the plan ... While dairy specialists were planning to collect 'surplus' milk from the pastoral herds, beef specialists were planning to increase production of slaughter cattle ... The Fulanis' own goals (eg. milk consumption, ensuring subsistence security), or the goals specifically of the women who sold milk (e.g. achieving more or less steady income despite fluctuating milk yields) were disregarded in the development planners' models...

In a sense, the introduction of dairy production among the Fulani was intended as a step to introduce them to the market economy. The traditional producers were overlooked when the Nigerian planners embarked on the modernization of the livestock sector. The end result is that Nigeria is today much more dependent on imported dairy products than ever before.

Urbanization, rural/urban migration and change of consumption patterns contributed to the increase in the consumption of milk and milk products. As a direct response to these planners, in both the Sudan and Nigeria, concentrated their efforts on commercializing the livestock industry and encouraging the private sector to invest in dairy farms around the principal towns alongside the river banks. This was accompanied by huge investments in feed factories, and the encouragement to businessmen to invest in drugstores and to import veterinary products. There is no reason to believe that this policy was successful in meeting the increasing demand for fresh milk. The milk produced by the private farms was too little, too expensive and not within reach of the poor who form the majority of the consumers.

Both Sudan and Nigeria encouraged the importation and reconstitution of powdered milk to solve the problem of milk shortages in the large urban centres. By this time the competition between the commercial sector and the small-scale producers became obvious. In the Sudan case, the private sector and reconstituted milk factories waged a war against the settled pastoralists urging the health authorities to investigate allegations of mixing milk with water, selling milk in dirty buckets and questioning whether the milk sold by pastoralists was not originally powdered milk. Consequently, the licensing system was reinforced and as a result many small producers were forced out of the market. Another problem was that the private dairy farms were competing over fodder with the small producers, and as a result fodder prices more than tripled during 1980–85.

A common feature of livestock development policies is their universality which, in effect, sacrifices variations in cultural traits and social objectives of production. They are generally dominated by the same principle which encourages the introduction of a modern input delivery system targeted to veterinary campaigns, water development, stock routes, commercial dairy farms, holding grounds and marketing facilities. This pattern of development in practice created a new system of social and economic differentiation in terms of access to credit, better grazing lands and water facilities. It is unfortunate that the capital intensive livestock industry system in the Sudan and Nigeria failed to fullfil its objectives and forced many small producers out of the market. However, it created vast skills in modern dairy farm and fattening grounds management, which are not accessible to the small producers who are still dependent on free range pastoralism.

PASTORAL DEVELOPMENT RELATED POLICIES

A common feature of pastoral policies in the Sudan and Nigeria is their advocacy since the late 1950s and early 1960s of pastoral settlement schemes sponsored by the FAO. In the Sudan case, the Project of Community Development for Settlement of Nomads in the Sudan was forwarded to the UN Agency for the Sedentarization of Nomads and Semi-nomads in 1962. An extract from the project document is relevant to the objectives of livestock development: first, it states that

> ... sedentarization is not an objective. It is only a means to improving the economic, social and cultural conditions of those communities, to integrate those communities into the life of the nation, and to enable them to contribute fully to national progress.

It is, therefore, assumed that pastoralists do not contribute fully to the national economy when they are on the move. This assumption is basically erroneous in that pastoralists contribute over 10–12 per cent of the GDP when they represent less than 15 per cent of the total population. Second, the project was not meant to encourage pastoralists to settle spontaneously with their cattle in the irrigated schemes. The mono-cultural agricultural policies adopted by the state did not cater for such an option. The project for settlement of nomads in the Sudan stated that they should be settled in the drylands instead. This point of view not only justifies the exclusion of pastoralists from integration into irrigated agriculture, but it also casts doubt about their ability to be cultivators. However, I cannot recall any pastoral community within the central rainlands of the Sudan which does not practice one form or another of crop production. Hence, the proposal to settle pastoralists in the drylands outside the irrigated schemes was meant to make available more land for the production of cash crops. The expansion of modern agriculture during the 1970s and 1980s, as described in the preceding discussion of agricultural policies, in practice, is clear evidence of the Government's intention.

The project for the settlement of nomads in the Sudan was criticized then by Asad, Cunnison and Hill (1962, reprinted in 1976). I quote their argument at length:

> If we ignore for the moment the extremely fundamental problem of getting nomads to accept settlement, it must be appreciated that the needs of settlement can only be achieved through investment that is sound from the point of view of the national community as a whole. Whether in the case of a particular group of nomads, such ends are best achieved through cultivation, ranching, dairy farming or some other industry, will depend on the potentialities of the area in which they live ... Similarly, the feasibility and desirability of settlement based on ranching will depend, among other things, on the capacity of other sectors of the economy to absorb the subsequent local labour surplus, or in other words, on the degree to which the economy as a whole can afford to eliminate traditional sources of productive employment. Unfortunately the plan shows little awareness of such problems.

The settlement of pastoralists was debated by policy makers and social scientists (summarized in Ahmed, 1976), and the only reason that it was not fully implemented by the Sudan Government was because the United Nations was somehow convinced that it was not a worthwhile effort. However, settlement as a policy was not completely abandoned and continues to dominate livestock development projects until today.

Similar developments were correspondingly taking place in Nigeria between the 1960s and the 1980s. The reasons behind the establishment of the grazing reserves is documented by several authors (Hopen, 1958; Awogbade, 1980). The objectives include livestock-agricultural integration and an intensive interaction with the market economy to increase off-take, while securing a reliable system for the provision of health, water and education facilities for the Fulani. The proponents of the grazing reserves (FAO, 1980; ILCA, 1979, 1986; Green Revolution, 1981) argue that the traditional systems of herd management represent a waste of labour and valuable land resources. Hence, the settlement of the Fulani in the grazing reserves is expected to make available vast stretches of land for cultivation and the production of food crops. An association between social development and settlement has, therefore, been perceived as the most pertinent way of developing pastoralists.

However, the opponents of grazing reserves (Awgobade, 1982; Okaiyto, 1982; Gefu, 1987, 1990) argue that the traditional systems of pastoral production are well suited to the savannah ecology.

Moreover, they assert that the Fulani have developed through the years an efficient system of resource management and that social development services can be provided to the Fulani while they are on the move. They have also rightly warned against the likelihood that the land appropriated from the traditional producers, both peasants and pastoralists, may be allocated to the already wealthy and powerful large-scale farmers for the production of cash crops to satisfy the demands created by the international market economy. Ecologically, the opponents of the grazing reserves argue that the concentration of livestock in limited reserves would ultimately contribute to over-grazing. In other words, the very facilities which have been provided for the development of the Fulani pastoralists would eventually lead to ecological imbalance within the reserves and in the areas surrounding them.

In the Sudan the pastoralists' demand for health, water and other social services forced the government to initiate development projects which became major election issues during the short periods of democratic rule which the Sudan has experienced. Due to the present economic recession, health facilities operate with minimum capacity or are not operational at all. Problems pertaining to the shortage of medicine and vaccines are common, and even when these are made available (locally by the National Veterinary Laboratory), the refrigerators are non-operational or there is no kerosene to operate them. The same applies to the impressive increase in modern boreholes with diesel engines which have suffered engine breakdowns and shortage of spare parts and fuel. Most of the reservoirs were silted as a result of poor maintenance and lack of fencing.

The period from 1983 to 1988 was a period of stress, great anxiety and uncertainty among pastoralists. It was characterized by a lack of sense of direction in Government policies, economic recession, drought and famine. Moreover, Government policies shifted emphasis from development to survival management. Pastoralists began to migrate to towns and relief centres in huge numbers. The war in Southern Sudan alone created over one million urban refugees (most of whom where pastoralists) living in apalling conditions, ravaged by starvation, disease and malnutrition.

It is unfortunate that the state was incapacitated and was not in a position to formulate a policy towards rehabilitation let alone development. Policy objectives such as herd re-building and raising productivity were not on the agenda. However, it is only fair to mention that export promotion measures constituted the main policy option available to the government. It was surprising to many observers that livestock exports increased during the 1983–1985 drought which created domestic meat shortages and forced the government in 1985/1986 to contemplate importing cheap red meat from Australia.

The Nigerian settlement schemes were also sponsored by the FAO, which in 1954 introduced the concept of creating grazing reserves for the Fulani pastoralists. The objectives of this policy are not different from those of the Sudanese project for the Settlement of Nomads.

The Nigerian case, on the other hand, illustrates how the planners define their objectives independent of each other which thus prompts contradictions in policies. Consider, for example, the estimates given by the Green Revolution (1981:31), which was designed for achieving self-sufficiency in food including livestock products. The planners' perception of increasing milk production from the traditional herds was based on manipulating livestock statistics like this: The average herd size in the traditional sector is 44 head per household. The mean milk off-take per day is estimated at 0.743 litre, a peak of 1.274 litres per day descending to 0.357 litres per day in the dry savannah. There are 9.3 million heads of cattle in Nigeria, out of which 3,208,500 are cows and of which 1,816,011 cows are in milk at any given time. The conclusion from this array of statistics is that, the traditional herd can provide 492,139 tons of fresh milk per annum at a farm-gate price of N 147 millions. This culture of figure manipulation is certainly out of touch with Fulani reality, where most of the milk is sold in the form of sour milk, *nono*, and the rest is consumed by the household. Waters-Bayer's (1988) study of dairy production by Fulani women illustrates that very little fresh milk is sold or consumed, and that most of the milk is sold in the form of *nono* (skimmed fermented milk) mixed with *fura* (spiced balls of cooked cereals). Moreover, whereas the planners are thinking in terms of the needs of the urban consumers for fresh milk and how to reduce dependence on imported powdered milk, the Fulani's cultural practice of selling sour milk is a product of their response to the consumers in the rural areas and their own level of technology, which may not allow them to preserve fresh milk for long periods of time. Nigeria's failure to achieve the targeted milk production from the national herd, certainly, has nothing to do with Fulani conservatism or failure to respond to the local markets. It is, by and large, a matter of lack of understanding of the Fulani reality which mitigates against the sale of one type of milk product rather than another.

When Nigerian policy formulations failed, livestock development planning during the late 1970s and the early 1980s shifted the emphasis from the provision of services for the grazing reserves, to the establishment of private dairy farms. This, in retrospect, means that the planners have once more resorted to imported cultural practices to develop dairy production. Exotic cattle breeds such as Friesians, Guernseys and Jerseys were imported from Europe, loans were extended to the private sector and research institutions and dairy farms were demarcated. It is obvious that the modern system of dairy production requires the importation of a whole industrial complex such as livestock feed mills, milking equipment, machinery for silage making, laboratories, expatriates, etc. By 1981 a number of animal feed mills were established with a total capacity of 600 thousand tons, and there was a mushrooming of commercial veterinary pharmacies to cater for the demands of the private dairy farms. A few years later the feed mills confronted the usual story of capacity under-utilization, spare-part problems due to hard currency rationing since the mid 1980s, shortage of raw material, etc.

In general, livestock development policies have not proved to be the ultimate solution to the problems confronting the livestock industry in Nigeria. The exotic imported cows encountered some serious problems and there were reports suggesting that, apart from the high cost of imported inputs, the performance of the imported dairy cattle, even under the best management practices with intensive health care, has been low. The low performance of the imported cattle is attributed in part to reproductive difficulties, low pregnancy rates due to poor organization of the breeding programme especially heat detection, long calving intervals, short lactation, poor management and little attention to balanced feeds in terms of concentrates (Green Revolution, 1981:54). Later reports asserted that the new system has suffered from:

1. high overhead costs,
2. the exotic imported cows failed to cope with the Nigerian climate,
3. the prices of the modern dairy products are too expensive and not accessible to the average consumers, let alone the poor strata of the population and
4. the owners of the dairy farms diverted the subsidies provided by the government to other activities such as commerce. As the economic recession began to hit Nigeria harder during the mid-1980s, many of the dairy farms ceased to operate while some incurred huge debts to the agricultural banks.

While the planners were overwhelmed by the passion to import exotic breeds, the settled Fulani were busy introducing traditional breeds to minimize the high loss of animals which they encountered during the first years of their settlement in the sub-humid zone where herds encountered different diseases, insects and natural hazards. The Fulani introduced Ketuku cattle, which have a high resistance to disease and insects, and crossbred them with Sokoto Gudali and White Fulani with positive results. Some households reported that their herds began to pick up and in a few years they could compensate for the losses which they had incurred during the previous years.

Another point of interest is the manner in which agriculture and livestock production are perceived as separate activities and the way plans to develop them are designed in isolation from each other.

According to Okaiyeto (1982:529), the River Basins Development Authorities and the Agricultural Development Projects have concentrated until 1978/79 only on crop production. This focus suggests that most of the Fulani traditional grazing lands are being transformed into arable lands. The end-result of this ill-conceived policy, as Okaiyeto puts it, is that, "the Fulani would then become marginal farmers at the best, and at the worst, landless labourers working for the governing, bureaucratic and other elites".

However, the issue is not one of a simple loss of land from one sector of agricultural production to another or conflicts over land resources. It is a complete loss of a valuable source of livelihood and an increasing inability to utilize the traditional methods of herd management and to handle natural crises such as drought and its accompanying shortage of grazing lands. The Fulani have lost one of the few ways of compensating for the loss in grazing lands during the difficult years.

The use of modern veterinary drugs which are, at times, not accessible to the pastoralists, may highlight the need for experimenting with traditional veterinary medicine. In this I share the view of Ibrahim *et al.* (1983:189) that

> ... the easiest and most rational solution to the problem is to develop acceptable inexpensive drugs from reasonably inexpensive sources for use as supplement to commercial drugs. Veterinary traditional medicine provides a shortcut to this end, and is more readily accessible for scientific investigation than is its counterpart, traditional human medicine.

This suggests that there is the possibility of using traditional veterinary medicine. This also implies that this valuable local knowledge has to be collected before it disappears.

THE RESPONSE OF PASTORALISTS TO THE CRISIS

The commercialization of pastoral production via livestock development policies has been met with different responses among pastoral societies. Large sections of the pastoral communities are integrated into the market economy, and their living standards are closely tied to the status of the national currency and the fluctuations of the prices of manufactured goods.

In general, Baggara, Kababish and Rufa'a pastoralists in the Sudan, and the Fulani of Nigeria have been displaced as a consequence of the expansion of agricultural schemes, and they have devised, with great unease, new migratory patterns to allow themselves longer periods in the southern sub-humid zone. Others settled and became more dependent on farming in order to avoid the purchase of grain. Those who migrated to the outskirts of towns have been transformed into milk sellers dependent on their small herds of goats and cows. They, unfortunately, have to compete with the modern sector, dairy farms and imported reconstituted milk. Herd management is practiced by women while most men are involved in firewood and charcoal selling or looking for jobs in towns in order to diversify sources of income and to make ends meet.

Pastoralists displaced by the 1983–85 famine have responded in various ways to the crisis. Pastoralists with small number of animals began to practice sedentary pastoralism in large villages, or migrated to live in squatter settlements around the large urban centres. Those who lost all their animals migrated to relief food distribution centres only to live as destitutes depending on international charity and relief food. Baxter depicts this trend as a direct result of the underestimation of the skills possessed by pastoralists which have aided them over the years. According to Baxter:

> Family life and core social and economic relationships have taken a beating. Pastoral men have grown accustomed to having their skills devalued, and have adapted to migrant labour and shanty town life; as their womenfolk have adapted to getting by on relief handouts while trying to maintain, if they are lucky, a heifer or two and a few sheep or goats. (1987:16)

Pastoral societies have different local level responses to development policies and these cannot be discussed at length in this paper. Although pastoralists are resilient to change, their social structures respond differently to crises. Rich pastoralists are much more able to diversify their incomes than poor pastoralists and different ecological conditions create different responses and survival strategies.

In the Fulani case the emergence of a number of veterinary nurses operating outside the official channels is a clear indication that pastoralists prefer to depend on their own experience with modern inputs. This case reveals that they can operate flexibly independent of the bureaucratic strings imposed by the official administrative structure. The Fulani response to the agents of change is impressive, despite considerable difficulties resulting from the new ecological and economic situation. This is not to argue that the Fulani have not suffered from the disastrous impacts of ill-planned policies. However, pastoralists in the Sudan were not able to devise such a system for many reasons, major ones are the state's restriction on drug imports and the ailing economy.

It is possible to elucidate two processes without which the pastoralists might not have been able to cope with the crisis. First, micro-responses related to tapping the indigenous knowledge at times of crises (collection of edible plants, cross-breeding of cattle varieties to reduce susceptibility to diseases, use of traditional medicine for animals and the mobilization of labour within and outside pastoral production). Second, at the macro-level, responses to increasing impoverishment which produced counter processes (flight and settlement in the sub-humid zone due to drought, increase in the prices of manufactured goods relative to those of cattle and milk, loss of land to mechanized schemes and the River Basin Authorities, as well as the state policies which favour agricultural production). At both levels pastoralists have to accommodate to the new situation with great unease. While wealthy households are able to diversify their career patterns, poor households are increasingly marginalized with grave consequences, such as accepting wage labour in agriculture or as herders, migration to towns and centres of employment. Scenes of the impoverished Fulani who joined the pool of the urban poor in towns such as Sokoto, Kano, Kaduna and Zaria are staggering by all standards, as are those in Omdurman, Khartoum, El-Obeid, Malakal, Wau, Kassala, and Port Sudan. Nevertheless, even in town, the pastoralists have to adapt themselves to a new sociocultural milieu in which their pastoral experience is of scanty, if any, relevance. With no skills to secure them jobs in the towns' modern sector, the pastoralists in town are among the most wretched on earth.

DISCUSSION AND CONCLUSIONS

An attempt has been made to discuss some of the issues pertaining to the differing impacts of livestock development and pastoral development policies on pastoral societies. In this case the available planning culture refers to modern knowledge and its associated attitudes which have been acquired through Western education systems *vis-à-vis* pastoral culture and society. There is a need to develop contact points between planners and pastoralists and to reconsider seriously the integration of pastoral and agricultural production, the use of browsing trees to supplement seasonal deficiences in pasture and the utilization of traditional veterinary medicine to supplement modern medicine. There is, unfortunately, some set-backs in these three areas emanating from the present planning culture. The planners have come to realize the importance of the integration of livestock and agricultural production after the large-scale mechanized schemes have imposed tremendous constraints on the Fulani pastoralists. However, these difficulties are surmountable and there is still the possibility of livestock agricultural integration in these schemes. For example, to include the production of legumes in the rotation system and crop by-products as fodder.

Further, the availability of medicinal plants has also been negatively affected by the indiscriminate cutting of trees and very few scientists are interested in investigating the usefulness of traditional veterinary medicine. It seems that the competition for the huge drug market in Nigeria will be won by the multi-national companies, with whom the governing elites have much more in common than with the Fulani. Again, the Fulani knowledge about their ecology could be used to develop new inexpensive drugs for their animals with the aim of supplementing the unobtainable and expensive imported animal drugs.

Settlement policies and grazing reserves have not succeeded in offering the pastoralists any of the modern inputs which they promised. The Livestock Service Centres which were established in the 1970s are still non-operational, without any drugs, boreholes or any other form of services. However the reserve is extremely useful in offering the Fulani the possibility to settle in areas in which they would have otherwise been denied the opportunity to utilize. It has, in a sense, provided an adequate answer to the land question in favour of the Fulani pastoralists. Nevertheless, it failed to offer them a viable alternative in terms of an efficient modern input delivery system. The problem is what does the future hold for the Fulani? If more are to settle in the same grazing reserve what are the chances for opening up new grazing reserves amidst an increasing population pressure in the northern states of Nigeria?

In conclusion, it is impossible to develop pastoralists without due consideration of their socio-cultural practices and the manner in which they respond to the agents of change. It is also naive to think that the knowledge of the pastoralists alone is a sufficient prescription for development. There is the need for modern knowledge which can penetrate areas of concern not attainable by the use of indigenous standards of living alone. This can be useful to advance the pastoralists' standard of living through the improvement of health, education, water supply and an increase in animal productivity. The pastoralists are very responsive to such useful interventions, as the informal veterinary medicine-men illustrate. Therefore, more cooperation between the technical and the social science perspectives is needed; while a serious appeal to indigenous knowledge may pave the way for a better understanding of some of the merits of the pastoralists' perception of ecology while exploring the value of adapted technological interventions.

As a cultural system, development planning carries new cultural meanings whose acceptance by recipients is a prerequisite to the success of such plans. Livestock development and pastoral production cultures are in confrontation: the former is geared towards modernization through the adoption of imported technology and institutions while, on the other hand, pastoralists' culture aims at securing the basic needs of the household. Whereas livestock development depends heavily on imported technology, knowledge and infrastructure, the latter is based on local knowledge with very little use of modern inputs. The contradiction between these two cultures of pastoral production are compounded by an interest gap between the pastoralists and the planners. The planners are educated and trained to appreciate modern techniques of production with little or no interest in Fulani cultural factors of production. These contradictions also relate to the planners' perception of the methods capable of developing the livestock industry, which are at times incompatible with the pastoralists' objectives and organization of production.

Contrary to this system of socio-economic and cultural production, the planners and administrators believe in the simplistic view that an increase in livestock production and productivity is synonymous with pastoral development. The prime objective of integrating pastoralists into the market economy is to satisfy the increasing demand for meat and other livestock products. This view, certainly, does not correspond with that of the pastoralists, whose perception of animal husbandry is geared towards satisfying the immediate needs of their households. It is in this respect that pastoral development policies failed to engineer development in harmony with pastoralist social objectives and physical environment. Livestock development policies have so far been mainly concerned with the provision of water, disease control and raising production with little interest in social development. Since development funds are allocated according to investment priorities, pastoral development has often been considered secondary to cash crop production and commercial ranching.

Abandoning pastoralists or leaving them to struggle alone for survival assuming that they will live in harmony with nature, is a misjudged objective. Positive interventions are urgent in order to improve the standards of living of pastoralists, and any delay in doing so will prolong their misery and render them victims to the increasing hegemony of the state apparatus, ecological pressures and to the livestock traders, middlemen and marketing boards. Hence, any call for conserving or romanticizing the present situation of the pastoralists is merciless, and would earn them nothing but more suffering and misery.

The issue here is not a simple increase of production for the markets *per se*, but a process which should be accompanied with a massive social development programme engineered by the pastoralists themselves, with their immediate needs as a priority. As facilitators we have better chances of success if we begin with small-scale interventions working together with pastoralists, and not compel them to accept the assumption that the experts know what the pastoralists need and how to get access to it. It is about time that pastoral knowledge be utilized and that development experts explore them in their efforts to generate sound pastoral development projects.

REFERENCES

Ahmed A.M., 1976: *Some Aspects of Pastoral Nomadism in the Sudan*, Khartoum. Khartoum University Press.

Asad T., I. Cunnison and L.G. Hill, 1976: "Settlement of Nomads in the Sudan: A critique of present day plans". In A. M. Ahmed (ed) *Some Aspects of Pastoral Nomadism in the Sudan*, Khartoum. Khartoum University Press.

Awogbade O.M., 1980: "Livestock Development and Range Use in Nigeria". In Galaty, J.G. *et al.* (eds) *The Future of Pastoral Peoples*, Ottowa. International Development Centre.

Awogbade O.M., 1983: *Fulani Pastoralists: Jos case study*, Zaria. Ahmadu Bello University Press.

Awogbade O.M. and S. Famoriyo, 1982: "Nomadic Land Use and Beef Production in Nigeria", in O.A. Osinowo *et al.* (eds), The *Proceedings of the Conference on Beef Production in Nigeria* (July, 1982) Ahmadu Bello University Press for the National Animal Production Research Institute (NAPRI).

Baxter P.T.W., 1985: "From Telling People to Listening to Them: Changes in approaches to the development of pastoral peoples". In *Manchester Discussion Papers in Development Studies*, International Development Centre, Faculty of Economic and Social Studies, University of Manchester.

Baxter P.T.W., 1987: "The New East African Pastoralism: An Overview". In *Edinburgh Anthropology*, 2:1–25.

FAO, 1980: *The Milk Market in Nigeria.* Brussels, Commissionn of the European Community.

Gefu J., 1987: "Land Use in Nomadic Pastoralism and the Issue of Sedentarization", in M. Mortimore *et al.* (eds), *Perspectives on Land Administration in Northern Nigeria*, Kano, Bayero University.

Gefu J., 1988: "Livestock Paradox in Nigeria: Insights from some development perspective". In L.A. Ega and T.K. Atala (eds) *Rural Development policies in Nigeria*, Proceedings of the 4th National Conference on the Nigerian Sociological Association, Zaria, Ahmadu Bello University.

Government of Nigeria, 1983: *Green Revolution*, Lagos.

Hopen C.E., 1958: *The Pastoral Fulbe Family in Gwandu*, London, Oxford University Press for the International African Institute.

Ibrahim M.A. *et al.*, 1983: "Traditional Concepts of Animal Disease and Treatment Among the Fulani Herdsmen in Kaduna State", Nigeria. ODI. *Pastoral Development Network Paper* No. 16.

ILCA, 1979: *Livestock Production in the Sub-humid Zone of West Africa*, Addis Ababa, International Livestock Centre for Africa.

Okaiyeto P.O., 1982: "Livestock Development Strategies in Nigeria", in O.A. Osinowo, *et al.* (eds), *The Proceedings of the National Conference of Beef Production in Nigeria* (Kaduna, July, 1982). Ahmadu Bello University Press for the National Animal Production Research Institute (NAPRI).

Raikes P., 1981: *Livestock Development Policy in East Africa,* Uppsala, Scandinavian Institute of African Studies.

Sandford S., 1983: "Better Livestock Policies for Africa". *ALPAN, Network* Paper No. 1. International Livestock Centre for Africa (ILCA).

Sandford S.,1983: *Management of Pastoral Development in the Third World.* John Wiley. Chichester.

Sudan Government, 1960: *Progress: 2nd Anniversary of Sudan Revolution,* General Information Office , Khartoum.

Sudan Government, 1961: *Progress: 3rd Anniversary of Sudan Revolution,* General Information Office, Khartoum.

Sudan Government, 1963: *Progress: 5th Anniversary of Sudan Revolution,* General Information Office, Khartoum.

Sudan Government, *Economic Survey,* Ministry of Economic and National Planning, Khartoum. Several issues 1960–1986.

Sudan Government, *Ten Year Plan, 1960–1970.*

Sudan Government, *Five Year Plan, 1970–1974.*

Sudan Government, *Six Year Plan, 1978–1984.*

Waters-Bayer A., 1988: *Dairying by Settled Fulani Agropastoralists in Central Nigeria,* Kiel. Wissenschafts Verlag. Vauk.

Indigenous Agro–Veterinary Knowledge System: An Alternative Resource

Mamman Aminu Ibrahim

There is a growing international realization, which is welcome, that science and technology are not culturally neutral, and that "modernization" is not necessarily synonymous with "westernization" (Prigogine, 1979; NIRA, 1988; Topel and Pesek, 1989; Elmandjra, 1990; FAO, 1990). Western agencies have at times made serious mistakes by trying to transplant technology developed in one culture, under one economic system, and suited to one set of climatic and soil variables, into a Third World country where all these factors are very different (Topel and Pesek, 1989). The FAO (1990) has stated bluntly and truthfully that solutions to pressing problems such as economic stagnation and environmental degradation have remained elusive despite almost half a century of development activities in the arid and semi-arid lands of Africa. This realization is generating a number of new areas of study.

Ethnoveterinary Research and Development (ERD) is one of the newest directions in ethnobiology (McCorkle, 1986). It is the systematic investigation and practical application of folk veterinary knowledge, theory and practice within a holistic but comparative and production system specific framework (McCorkle, 1989).

It is convenient to discuss ERD under the following sub-themes:

1. *Veterinary Ethnosemantics and Ethnotoxonomy* deals with the most fundamental concepts of ERD, namely indigenous veterinary and husbandry concepts and science, and the relationship between folk and modern scientific taxonomies.
2. *Ethnoveterinary Pharmacology and Toxicology* investigates folk uses of plants and other materials in the control of animal diseases. The importance of this area lies in providing culturally

appropriate, economically feasible and consistently available drugs of natural origin.

3. *Ethnoveterinary Manipulative Techniques* studies and improves indigenous methods of vaccination as well as surgical procedures such as castration and bone-setting.
4. *Ethnoveterinary Magico-Religious Practicies* catalogues and analyses religio-cultural profiles insofar as these have bearing on animal production and health, for instance concepts relating to diseases whose aetiologies are not visible to the naked eye.
5. *Ethnoveterinary Education,* hitherto not formally admitted into ERD, should study how the foregoing areas continuously evolve and how they are transmitted across generations, since this may guide the introduction of interventions.

This paper discusses the potentials of Ethnoagro-veterinary knowledge as a resource for human, and particularly rural, development in Nigeria, and Africa in general, drawing upon studies conducted in the nation's arid and semi-arid zone.

ERD AS A RESOURCE FOR HUMAN DEVELOPMENT

It is presumptuous to attempt to improve a long-existing system without first thoroughly understanding it within its own context. For this reason, there is hardly an aspect of development that will not benefit greatly from systematic study of Indigenous Knowledge Systems (IKS), of which ethnoveterinary research is a part. Here, only a few areas will be highlighted, including transfer of technology, formulation of indigenous research policies, education, as well as animal production and health.

"TECHNOGRAFTING"

Developing countries spend most of their resources attempting to graft exotic technologies, and the rest of their resources trying to maintain the technografts in place, and all their time bemoaning the outcomes. They are assisted in this futile venture by both the well-meaning and the unscrupulous alike. The effect so far is feelings of bewilderment and futility and despair.

To maintain a mismatched tissue-graft even temporarily in place is generally recognized to be both medically difficult and foolhardy, regardless of the desirability for the operation. We owe this medical understanding to advances in the science of immunology. Some similarities between cultural and biological evolution have been shown by several anthropologists, and were popularized by Skinner (1971). It may be that similar analogies can be drawn between technology transfer and tissue-grafting. Perhaps anthropology can do for Development Science what Immunology has done for Medical Science. Any given technology comes wrapped in cultural, environmental and other assumptions. A tractor presupposes a small rural population, large-scale farms, the right terrain, and maintenance services; an appropriately labelled insecticide assumes a literate user and specified storage facilities; a 'slaughter-clearance period' for persistent drugs assumes a population more at risk of suffering from the effects of consuming drug-residues in animal products than from starvation or death.

The term "technografting" is used here rather than e.g. "transfer of technology", in order to underscore the need to begin typing and matching technologies and cultures prior to their transfer, to emphasize the consequences of mismatches, and as a continual reminder of the seriousness of transfer endeavours. It may also suggest a theoretical framework through which to understand some development problems, which can sometimes be better understood as "technograft-rejection syndromes". For example, lack of knowledge of local beliefs, practices, and prejudices has long been recognized to be one of the factors complicating successful adoption of modern poultry production in developing countries (McArdle, 1972). Among the Hausa and Fulani in the semi-arid zone of Nigeria there is probably no better demonstration of this than the people's belief that hens have invisible breasts from which chicks suckle invisible milk (Ibrahim and Abdu, 1990).

During the last two decades, it has been thought desirable to upgrade genetically the local chicken stock of subsistence rural farmers in developing countries using a procedure referred to as "Operation Coq" (Taran, 1974), and which had been used with some success in Israel and India. In Nigeria all the local cockerels in co-operating villages were to be substituted with exotic ones within one or two years (AERLS, 1976). In addition, day-old chicks of improved breeds were imported for distribution and sale to rural farmers.

These rural development programmes sound reasonable. Unfortunately, the experts did not know that rural Hausa and Fulani classify poultry as mammals. It is the experience of the people that animals which aggressively care for their young invariably have mammary glands for nursing, and they are loathe to make an exemption for the hen. The people thus consider it to be wicked to procure "mother-less" day-old chicks. This belief is present even among some staff employed in modern poultry farms, who explain to themselves that the ritualistic work which they have to perform on the farms is white-man's magic.

On the other hand, the rural Hausa and Fulani have sophisticated and exact empirical knowledge of egg-incubation periods, and are able to hatch eggs of different poultry (chicken, ducks, and guinea-fowls) simultaneously by a foster arrangement. The rural improvement scheme might have succeeded better if fertilized eggs had been offered rather than 'mother-less' day-old chicks.

"Operation Coq" failed because it contradicted the people's agro-veterinary knowledge and experience. First, the people cull cockerels when they observe the small, yolkless, so called 'pigeon-eggs' in a hen's nest, since they believe such eggs have been laid by aged cockerels (Ibrahim and Abdu, 1990). "Pigeon-eggs" are not uncommon, and the introduced cockerels which are bigger were more likely to be culled than the small local ones. Next, no attention was given to introducing cockerels which had culturally-favoured feather colours. Also, prevailing management style, diseases, level of veterinary care, and the tropical climate were not suited to birds which had been bred for temperate and "high-tech" environments.

In other cases, actual severe damage is caused when alien concepts are introduced even with good intentions. New things have to be related to existing ones by the associative intellect. Where the new and the old are too far apart, a mistaken association is forced, sometimes with disastrous results. An example is the application of local concepts of indigenous drugs to modern pharmaceuticals. The Hausa have a multi-dimensional concept of plant use (Etkin and Ross, 1982), such that most plants used as drugs are also used as food for both animals and man. Indeed, even plants used as insecticides may also be used as ingredients in food (Ibrahim, 1990a).

These concepts, which are informed by experience with botanicals of low potency, toxicity and persistence, are unfortunately applied also to the vastly more potent, toxic and persistent synthetic drugs being imported. The impact and significance of this to human and animal health are enormous. Aerosol insecticides are sometimes sprayed on meat and fish in open markets to keep away flies. Fishermen at times improve their catch by applying DDT in dams which also supply drinking water to towns.

INDIGENOUS RESEARCH

There are good reasons why developing countries should place more emphasis on indigenous research (Aliu and Mohammed, 1990). But, as Abdus Salam (1989) has pointed out, an important stage in this development should be awareness on the part of indigenous scientists of the problems which plague their own countries, and which most preoccupy their own people. It is presumptuous to suppose that indigenous research will naturally take its cue from within. Indirect evidence has been published (Bozimo 1983) which suggests that Nigerian scientists take their research cues from elsewhere. This was shown by very low impact factors calculated in respect to a Nigerian Journal. Out of a total of 1,303 citations in the journal during a period of 10 years, only 17 (1.3 per cent) were to the Journal itself, and these were in fact mainly self-citations. More direct evidence can be provided by analysing data on who requests for articles published by Nigerian scientists in various journals. Such a study is in progress (Ibrahim, 1990b). It can be seen from Table 1 that, of 865 requests received from 53 countries in respect of articles by staff of the Ahmadu Bello University's Faculty of Veterinary Medicine, only 0.7 per cent were from West Africa. It is indeed very clear from Table 1 that veterinary scientists from developing countries with similar problems do not request published information from each other and, therefore, can not take cues from each other.

For indigenous research to make an impact where it is needed most, it is necessary for a sufficient number of the researchers to be working on the same indigenous problems in a systematic manner. It is difficult to see how this can be achieved without study of both the limitations and useful aspects of indigenous knowledge and practices. Hitherto this has not been evident. For example, it seems to be that the objective of genetic research aimed at improving livestock species in Nigeria is to produce stock considered "good" by western standards. Most Nigerians prefer guinea-fowl eggs and other hard-shelled eggs to hen's eggs, they taste 'better' and survive rough rural transport (Ayeni and Ayanda, 1982; Ayorinde, 1988), and which are used in the rural game of 'fashe' (Ibrahim and Abdu, 1990). Indigenous research objectives should aim at improving the egg-laying ability of the domestic guinea-fowl rather than the domestic chicken. Furthermore, to the rural dweller in semi-arid Nigeria, the performance of a chicken lies in her total ability as a mother. This is based on the number of chicks she successfully weans. Clearly, meaningful indigenous research should not base selection and breeding solely on such western standards of "performance" as egg size and number under cage conditions.

CUSTOMIZING EDUCATIONAL CURRICULA

It ought not still to be a subject for debate whether or not culture plays any role in the learning process in all disciplines, (Mu'Azu, 1980), since all learning is a modification of some prior knowledge (Popper, 1977). Some of the problems of applying formal western education in African countries have been reported (Rayfield, 1983; Friere, 1978) and reviewed by FAO (1990). Western education in general, and science education in particular, in the form inherited from colonial administrations lacks a proper home-base in Africa. Most science students in Nigeria are at a loss to relate the acquired "education" to their immediate environments (Mu'Azu, 1980). Apart from alienation, little has been achieved by teaching sub-atomic physics in its western form directly to students who return home to firewood technology and a universe where the rains fail owing to the wicked arts of the village prostitute (Ibrahim and Mu'Azu, 1990).

The impact of 'modern' education in Nigeria after almost 30 years of independence is as follows. An accountant may purchase a chicken at the cost of a ram based on a "money-doubling" prescription from a witch-doctor; a professor of pathology may pay 'rain-makers' to suspend the rains to allow a ceremony to take place (Ibrahim and Abdu, 1990). Magical secret societies which engage in brutal magico-ritual ceremonies are common in Institutions of Higher Learning, including Universities of Science and Technology and Polytechnics, and several students have been expelled or arrested on this account. Items such as these are now commonly reported in the nations electronic and print media, for example, *New Nigerian* No. 7506 of 2nd Feb, 1990: p. 3; *Triumph* No. 1335 of 1st March, 1990:12.

The most important contribution of ethnoveterinary research and other components of indigenous knowledge may well be the provision of the cultural knowledge-base on which to design effective science educational curricula. Until this is done, science education will continue to be totally unrelated to people's beliefs and environments. It must be recognized, after all the time and investments lost, that it is possible to memorize or to do some aspects of science ritualistically, without any real understanding of the scientific method. If a few should really become educated, this would be in spite of the present system rather than because of it.

There are at least three ways in which ethnoveterinary and other components of indigenous knowledge research may serve educational purposes.

Customizing rural school environments

In arid and semi-arid Africa traditional education has four characteristics (Rayfield, 1983; FAO, 1990): it is completely effective; every child who survives it is allowed to "graduate"; its costs are not prohibitive; and children are not withdrawn from the work force. All these are lacking in the present educational system, partly explaining its failures in the rural areas. A study of traditional methods of training youths may assist in finding workable school environments and time-tables.

Breaking protection-codes for unfounded beliefs

Many unfounded beliefs are protected from rational enquiry, which is why they survive among literate Nigerians. The protection devices are normally hidden, and require patient research to expose them. For example, the protection for the Hausa belief about hens having mammary glands is as follows (Ibrahim and Abdu, 1990): to see the breasts of a chicken one has to first disturb a hen that is sitting on her chicks, since the breasts emerge only when the chicks are suckling. Also, one must use either the leaves of *Anona senegalensis* or a circular mat (*faifai*) that had previously been used to collect grain flour. Before trying the experiment, one must also be prepared to either die or to become insane, unless one possesses certain protective magical charms which are known only to witch-doctors. Should one try the experiment exactly as prescribed and still fail to demonstrate breasts on a hen, this is not considered sufficient evidence against the belief. It is held that some chickens do not have mammary glands, such are in fact supplied by witch-doctors, at great price for use in powerful magical prescriptions. Perhaps, by chance, one used such a rare chicken for the experiment. Naturally, not many people are willing to try the prescribed experiment!

The Hausa and Fulani also believe that butchers are really dogs in disguise. One may prove this by looking at their stand on market days through the eye of a hoe. Alternatively, one may marry the daughter of a butcher and then place a calabash under her bed at night, which should become filled with blood by morning. But to carry out either test, however, one should be prepared to either die or become insane.

Customizing contents of science curricula

A substantial proportion of the peoples' knowledge relate to animals, diseases, and medicinal herbs, as well as to the environment. These can serve as sources of materials for scientific illustration and teaching. This holds even for outrageous claims. For example, the vicious female evil spirit called Inna by the Hausa is the mother of Bawa Makau, another evil spirit. She is believed to drain away a victim's bone-marrow, causing paralysis and muscular atrophy, a condition called shan Inna. This is an acute anterior poliomyelitis, and the Inna tribe of spirits thus consists of the polio viruses (Ibrahim, 1990a). There is no reason why such associations should not be made in teaching or incorporated into curricula.

Gajimari is perhaps the most feared evil spirit among the Hausa in Nigeria's semi-arid and sub-humid zone. An investigation into the nature of this "spirit" is in progress (Ibrahim and Mu'Azu, 1990). A combination of common and unusual natural phenomena, exact observations, and preoccupation with climatic and environmental problems have generated a myth which relates the causes of sporadic deaths of humans and animals to rainbows and why some clouds fail to release their rain. The Gajimari myth provides a source of spectacular richness for teaching integrated science and the scientific method at almost no cost in terms of "high-tech" laboratory equipments. In addition, the vast folk knowledge involving animals and medicinal plants, as well as classification systems, can be used for teaching and thereby help to bridge the yawning gap between the school and the real world, and ensure that useful components of indigenous knowledge are not forgotten.

There is nothing to lose and everything to gain by exploiting indigenous knowlege resources to enhance the development process. Certainly, no "loss of face" is involved. Even in the 1940s there were Englishmen who believed that a crock of gold was to be found at the foot of every rainbow (Minnaert, 1959). According to Bertrand Russell (1952), it was the spread of the scientific outlook, which began from the time of Charles II, rather than any definite argument which served as the solvent of unfounded beliefs in Britain. But, in African societies, economic growth has not benefited greatly from research or technological invention (Aliu and Mohammed, 1990). We may have to use more definite arguments to spread a sufficiently scientific outlook.

IMPROVEMENT OF ANIMAL HEALTH AND FOOD PRODUCTION

Ethnoveterinary and ethnoagricultural research have direct application in improvement of animal health and food production. The present economic state of most of sub-Saharan Africa has rendered the cost of veterinary care prohibitively high, and most recommended drugs are not obtainable at all; where they are available, cases of misuse are common owing to cultural and other factors. Drug companies are loathe to invest in drug development for a relatively small and poor market and for diseases that are peculiar to the sub-region.

The simplest and most rational solution to these problems is to study systematically indigenous materials and methods of treating the diseases, selecting potentially useful ones for development into drugs (Ibrahim, 1986). The value of such research lies in the provision of improved folk pharmacotherapy which is culturally appropriate, economically feasible and consistently available (McCorkle, 1986). Ultimately such research may also contribute new chemical leads for "high-tech" synthesis, but there is no reason why improved crude drugs which are proved to be reasonably safe and effective should not be used to help solve pressing problems.

There are more than 300 plants used in traditional control of helminthic infections of man and his animals in Africa (Ibrahim, 1984); a further 100 or more are used as insecticides for food protection and control of ectoparasitic infections in the west African sub-region. Hundreds of plants are similarly used for microbial and other diseases. Although only a few of these plants have been investigated, the indications are that many of them may be useful therapeutic agents. Of thirty plants tested for anthelmintic action, six were acceptably active for treatment of hookworm infections; the others may well be useful for treating other endoparasitic infections of livestock (Ibrahim *et al.*, 1984). In Cameroon, some traditional crop protection botanicals have recently been shown to be effective (Schrimpf and Dziekan, 1990).

It is important to recognize that the objective of most ethnoveterinary therapies is to achieve a clinical improvement of the patient rather than to eliminate causative agents, unlike in orthodox veterinary practice. Consequently, research protocols aimed at validating claims have to take this into account. For example, *Adansonia digitata*, used to treat diarrhoea in calves, may not kill worms or bacteria in laboratory experiments; but it is very rich in Vitamin C, which is an antioxidant vitamin now known to enhance immunity of animals to diseases (Bendich, 1987). In other cases, information is often available regarding their pharmacological efficacy. A mental-block which insists on the necessity for using pure compounds, even where these are not available, continues to ensure that Africa's useful crude plant drugs remain in laboratories in the form of "on-going" projects, because of the non-availability of equipments and indigenous drug companies. Meanwhile, huge sums of money continue to be lost to livestock diseases, and the local farmers continue to use their plant drugs without benefit of local scientific guidance.

Just as for drugs, evidence is increasingly being furnished by scientists for the rationality of indigenous management practices in Africa which, hitherto, have been considered to be erroneous. For instance, it is thought that African herders deprive young calves of milk by selling some of it for human consumption, and that this surely must reduce weight gain of the calves. However, a recent study of patterns of cattle growth in Kenya shows that the traditional herders' production system is quite efficient, providing milk for humans while producing viable young stock that eventually attain weights equal to those of calves fed all the milk produced by their dams (Coppack, 1990).

PROBLEMS OF ETHNOAGRO-VETERINARY RESEARCH IN NIGERIA

Apart from the usual problems of insufficient funding and facilities, there are some problems which are peculiar to ethnoagro-veterinary research in Nigeria, and perhaps other areas in Africa. It is a new field, and there are not enough scientists working in it. Indeed, it is a fact that most of the initiative for studying indigenous knowledge systems is either from western countries or from international development organizations. Secondly, it is an area of research that requires a multi-disciplinary and team approach to research. It cuts across disciplines, as indeed all real problems do. For, as has been said by Marvin Harris (1978), nothing in nature is quite so separate as two mounds of expertise! Unfortunately, this calls for a high level of funding and coordination of research efforts which are not presently available in African countries.

The reservoirs of indigenous knowledge resources tend to be the elder generation of rural dwellers. Their numbers are reducing as the younger ones leave for the towns with only half-formed beliefs and without going through the traditional educational training. Systematic investigations and documentation have to be done before it becomes too late (Nwude and Ibrahim, 1980).

In addition, it involves the collation, cataloguing and analysing of vast amounts of information. The information tools required are either non-existent, inadequate or non-appropriate for developing countries. Some advanced countries and organizations have established computer data-bases on aspects of medicinal plant research, but these are not suitable because they are too specialized and the target populations have no access to them.

RECOMMENDATIONS AND CONCLUSIONS

There is enough evidence to warrant calling for more emphasis on ethnoagro-veterinary and other indigenous knowledge systems research. Assistance is required from more advanced countries and organizations, in terms of funding, documentation and research facilities, and coordination of research efforts. Standardization and coordination of research protocols are particularly important in order not to romanticize the endeavour (FAO, 1990) and to validate experimental results (Ibrahim *et al.* 1983; Ibrahim, 1986).

Awareness needs to be generated, especially in the developing countries themselves. The department of Veterinary Physiology and Pharmacology at Ahmadu Bello University has established a medicinal and poisonous plant garden for exhibitions, and has this year introduced an elective course on Ethnoveterinary Medicine which is being taught to third year veterinary, science and sociology students. Such facilities can be improved upon and used for teaching and research, especially by students and staff from advanced countries who are interested in International Agriculture, Development Science, and to foster international relations.

Table 1. *Regional distribution of requests for published Nigerian veterinary information*

Source of request	Number of countries requesting reprints	Reprint requests	
		Number	% of total
North America	2	300	34.7
Western Europe	14	195	22.5
Eastern Europe	7	143	16.5
South America	9	89	10.3
Asia	5	46	5.3
Southern Africa	4	29	3.4
Australasia	4	28	3.2
East Africa	2	25	2.9
West Africa	3	6	0.7
North Africa	2	3	0.4
Central Africa	1	1	0.1
TOTAL	53	865	100.0

Source: Faculty of Veterinary Medicine, Ahmadu Bello University, Zaria, Nigeria

It is clear from the foregoing that optimal use of indigenous agro-veterinary knowledge will require the teaming-up of staff and students from such diverse fields as basic sciences, agriculture, veterinary medicine, education, anthropology, sociology, extension, linguistics, as well as environmental, development and information sciences. Such an endeavour may be costly, but not as costly as the alternatives that have already been tried with unsatisfactory results.

REFERENCES

AERLS (Agricultural Extension and Research Liason Services), 1976: "How to improve the performance of local chickens". *Extension Guide* No. 79. Ahmadu Bello University, Zaria:1–7.

Aliu Y.O. and A.N. Mohammed, 1990: "Indigenous research: a vital link between science, technology and development". *Discovery and Innovation,* 2(1):11–17.

Ayeni J.S.O. and J.O. Ayanda, 1982: "Studies of the husbandry practices and social acceptance of Guineafowl in Nigeria". *Bulletin of Animal Health and Production in Africa*, 30(2):139–148.

Ayorinde K.L., 1988: "Tips on backyard Guineafowl production in Nigeria". *Nigerian Livestock Farmer*, 8(2–4):10–12.

Bendich A., 1987: *Role of antioxidant vitamins on immune function. Proceedings of ROCHE Technical Symposium*, Daytona Beach, USA: March 11, 1987.

Bozimo D.O., 1983: "Nigerian scientific serial literature". *International Library Science Review*, 15:49–60.

Coppack D.L., 1990: "The bigger the calf the better the cow—true or false?". *Spore*, 26:11.

Elmandjra M., 1990: "Fusion of science and culture: key to the 21st century". *Discovery and Innovation*, 2(1):11–17.

Etkin N.L. and P.J. Ross, 1982: "Food as medicine and medicine as food". *Social Science Review*, 16:1559–1573.

FAO (Food and Agriculture Organization), 1990: "Herders' Decision Making in Natural Resources Management in Arid and Semi-arid Africa". *Community Forestry Note* No. 4:1, 67–69.

Frierre P., 1978: "Pedagogy in progress". Seabury Press. Quoted in FAO (1990).

Harris M., 1974: *Cows, Pigs, Wars and Witches*. New York: Vintage Books: vii.

Ibrahim M. A., 1984: *Evaluation of the Activities of Some African Traditional Anthelmintic Herbs for Activity Against Nippostrongylus braziliensis in Rats.* Thesis, Ahmadu Bello University, Zaria.

Ibrahim M. A., 1986: "Veterinary traditional practices in Nigeria". In R. von Kaufmann, S. Chater and R. Blench, (eds). *Livestock Systems Research in Nigeria's Sub-humid Zone.* Addis Ababa: International Livestock Centre for Africa:189–203.

Ibrahim M. A., 1990: "Toxicology according to Nigerian Hausa and Fulani agro-pastoralists". In C.M. McCorkle, E. Mathias-Mundy, and T.W.S. van Veen, (eds). *Ethnoveterinary Research and Development*. London. Kegan Paul International (in press).

Ibrahim M. A., 1991: "The Information Drain". *Development Policy Review*, 9:99–106.

Ibrahim M. A. and P. A. Abdu, 1990: "Ethnoagro-veterinary perspectives on poultry production in Nigeria". In C.M. McCorkle, E. Mathias-Mundy, and T. W. S. van Veen, (eds) *Ethnoveterinary Research and Development*. London: Kegan Paul International (in press).

Ibrahim M. A., N. Nwude, Y. O. Aliu and R. A. Ogunsusi, 1983: "Traditional concepts of disease and treatment among Fulani herders in Kaduna State, Nigeria". ODI *Pastoral Development Network* Paper 16(c):1–6.

Ibrahim M. A., N. Nwude, R.A. Ogunsusi and Y.O. Aliu, 1984: "Screening of West African plants for anthelmintic activity". *ILCA Bulletin*, 17:19–23.

Ibrahim M. A. and S. Mu'Azu, 1990: *A study of Gajimari, the killer rainbow spirit of Hausaland* (in preparation).

McArdle A. A., 1972: "Methods of poultry production in a developing area". *World Animal Review*, 2:28–32.

McCorkle C. M., 1986: "An introduction to Ethnoveterinary Research and Development". *Journal of Ethnobiology*, 6(1):129–149.

McCorkle C. M., 1989: "Veterinary anthropology". *Human Organization*, 48(2):156–162.

Minnaert M., 1959: *Light and Colour in the Open Air*. G. Bell and Sons Ltd:170.

Mu'Azu S., 1980: "Culture and the natural sciences". Paper presented to *National Seminar on Culture and Technology*, Kaduna, Nigeria: Sept 1–5.

NIRA (Japanese Institute for Research Advancement), 1988: *Research Output Agenda for Japan in the 1990s*. Quoted in Elmandjra (1990).

Nwude N. and M. A. Ibrahim, 1980: "Plants used in traditional veterinary medical practices in Nigeria". *Journal of Pharmacology and Therapeutics*, 3:261–273.

Popper K. R., 1977: "On Hypotheses". In P. N. Johnson-Laird and P. C. Wilson, (eds). *Thinking*. Cambridge University Press, London: 615.

Prigogine I., 1979: *The New Alliance*. Quoted in Elmandjra (1990).

Rayfield J. R., 1983: "The legacy of colonialism in education in West Africa". *Mimeo*, York University. Quoted in FAO (1990).

Russell B., 1952: *The Impact of Science on Society*. Unwin Paperbacks, London: 15.

Salam A., 1989: "Four stages of science transfer". *Whydah*, 1(6):11–12.

Schrimpf B. and I. Dziekan, 1990: "Tapping farmers' knowledge of crop protection". *Spore*, 26:12.

Skinner B. F., 1973: *Beyond Freedom and Dignity*. Penguin. London.

Taran M., 1974: "The development of the poultry industry in Israel and poultry improvement schemes in developing countries". *World Animal Review*, 10:38–43.

Topel D. and J. Pesek, 1989: "Indigenous knowledge and alternative agriculture". *CIKARD News*, 1(4):1.

Alternative Policies and Models for Arid and Semi–Arid Lands in Kenya

George A. Keya

Arid and semi-arid lands, here referred to as ASALs, comprise about 80 per cent of Kenya's total land surface area, support well over 25 per cent of the human population and slightly over half of the livestock population. A majority of the inhabitants are pastoralists although agro-pastoral and farming communities do exist. A prominent feature of the ecosystem is the low and variable rainfall, which rarely exceeds 800 mm, with most areas receiving 200–350 mm annually. The water resources are limited and poorly distributed. There are few permanent rivers in both the northern and southern rangelands of Kenya. Most of the other riverine catchments are characterized by seasonal streams that flow only during the wet season and remain dry for the rest of the year. Several boreholes and wells have been developed to tap the ground water resource.

This paper discusses Kenya Government Policies for the development of her arid and semi-arid lands with an emphasis on pastoral systems. The thesis is that inappropriate (or lack of) policies and models prescribed by experts have not had the desired success, and are in many ways responsible for the environmental crisis in the arid lands. Alternative approaches are, therefore, suggested.

A CHANGING SCENARIO

Although pastoralism has evolved over millennia and is, in many ways, a successful adaptation to the dry ecologically fragile arid ecosystem, recent interventions and trends in ASALs appear to militate against its continued survival. Sedentarization of the nomads has been encouraged through the provision of social amenities such as schools, market centres and hospitals. Frequent droughts have also necessitated settlement at famine relief centres. Inadequate security, due to inter-tribal raids in some areas, has also contributed to this trend of settlement. For example, of the 20,000 sq km area of south-western Marsabit district of northern Kenya inhabited by Rendille nomads, slightly over one per cent is man-made desert around permanent settlements. At the same time 40 per cent of the area is totally unused because of fears of inter-tribal raiding (Lamprey and Yussuf, 1981) and scarcity of water. The settlements have become centres of localized overgrazing and tree destruction. Development of permanent water sources in such human settlements has created permanently used rangelands, which have been so intensively utilized by livestock that they have become highly degraded landscapes devoid of vegetation cover.

In some areas loss of the herb layer, due to overgrazing and trampling, is followed by a steady decline of the woodland and shrub populations, as the pastoralists are obliged to move further every year to collect fuelwood and thorn-tree materials required for construction of livestock enclosures. Lamprey (1981) estimated that between six and ten trees are felled by each family of six people at each nomadic move, and that 50,000 trees are destroyed annually for *boma* (village, camp or animal pen) construction in the 20,000 square kilometres of Rendille country in northern Kenya in which the human population is about 30,000 people. This is an astounding revelation, not least because woodlands provide browse for the camels and goats, particularly during the dry seasons. The settlements have thus become centres of man-made deserts spreading out in concentric circles, as is evident in the Rendille settlements of Korr and Kargi.

The institution of administrative boundaries, wildlife reserves, forest gazettment, and the imminent encroachment of farming on arid lands due to population pressure has curtailed the migratory range of the pastoralists and their herds. Consequently, pastoralists are now returning too soon to previous camp sites before vegetation has had time to recover. A cycle of repeated forage depletion and land degradation is therefore set in motion. The effect of loss of land productivity on livestock performance is mostly manifest during dry seasons when feed limitations lead to high mortality.

It has been estimated that approximately 30–40 per cent of Kenya's arid and semi-arid regions are quickly degrading and that another 2 per cent has been completely lost to this process (NEHSS, 1984). In a country where only 1/3 of the land is arable and other natural resource endowment are generally lacking, this loss in productivity must be halted if sustainable production is to be achieved. A factor to be considered in this scenario is what appears to be a net migration of adults from the pastoral nomadic system to lowly paid jobs in cities and big towns. Young children, who would otherwise provide herding labour, are removed from the nomadic system to attend formal schooling. Although the implications of such migrations on labour requirements for herding duties have not been fully investigated, there is a general feeling among pastoralists that a labour shortage within the system is beginning to be felt. The ecological implication of this fact is that animals may never be herded far enough from home, thus leading to increased grazing pressure in vicinities of settlements. The consequences of such a grazing system have already been discussed.

Another feature of the pastoral system in Kenya is a changed diet among pastoral households. Pastoralists are consuming more grains and sugar than ever before, although figures on the actual magnitude are not readily available. This shift to grain cereals has occurred largely through the influence of famine relief provisions during drought periods. It must be noted that there has always been barter in this commodity (albeit at low levels in the past) between pastoralists and agropastoralists (Hjort, 1988). Although a fair exchange rate of livestock to farm products is vital for trade, the yearly increases in grain prices by the government and the relatively stable prices of beef seem to indicate that the livestock grain exchange ratio is lopsided in favour of grains, and therefore the pastoralists may be getting a raw deal for their livestock.

It is evident from the foregoing that pastoralists have been caught up in a web of development activities and consequences of which they have little understanding and control. The dilemma that faces the pastoralists in Kenya is, therefore, how to cope with a changed production environment that has been dictated by exogenous factors, while maintaining their traditional way of life. On the other hand, a challenge that confronts policy makers, planners, researchers and other development agencies towards a realization of the full potential of ASALs, is how to achieve an improved and sustainable production from the drylands.

DEVELOPMENT POLICIES

Policies for the development of ASALs began to emerge during the 1940s, when population increases in the African reserves were causing serious overgrazing, denudation and soil erosion. In 1946, a ten year development plan (1948–1958) was adopted to effect an agrarian change in pastoral areas (Mwangata, 1986). This was followed by the Swynnerton Plan (1954–1959) the policies of which were to be implemented by the African Land Development Board (ALDEV) through grazing schemes. The Range Management Division took over the function from ALDEV as an instrument of change in the pastoral economy at Independence, a responsibility which it continues to hold.

Although such policies existed development of ASALs was given a low priority throughout the colonial period, a situation that continued during the first three post-independence development plan periods. However, by the late 1970s, the government made a major evaluation of the potential contributions that ASAL areas could make to the national economy. Following from this evaluation a policy re-orientation recognized the importance of these areas in national development and the need to accord them special attention, considering that:

1. They have substantial potential for development though at higher costs than the rest of the country.
2. Most of the poorer people live in these areas, hence the need to improve their living conditions through increased productivity and the creation of employment opportunities that would enable them to share equitably the benefits of development.
3. The increasing problems of soil erosion, environmental degradation, the threat of desertification and the negative consequences of phenomena such as hunger and malnutrition often direct public resources to famine relief operations.

Kenya's formal development approach in the ASAL regions therefore began in 1979 during the implementation of the Marginal Lands Pre-Investment Study Project. Integrated projects, mainly agro-based, were launched with the help of donor assistance in some districts, with varying degrees of success. ASALs development was a donor specific, district based approach (Odinga, 1989):

> The individual donors set their priorities and operated in patterns often outside normal government administrative and budgetary system. Expatriate advisors played key roles at the district level. Useful experience from these programmes was not given wide application to improve the approach to ASALs.

The objectives of the first generation of ASAL development projects were human resource development, exploitation of productive potential, resource conservation and integration into the national economy. Not all ASAL regions have benefited from ten years of the ASAL development programme because of donor preferences and funding, which have not brought the funding of the programmes under direct co-ordination of the government.

Recognizing that ASAL development requires restructuring, the government of Kenya has outlined the main objectives of the second generation of ASAL development policies in the current National Development Plan (1989–1993) as:

1. Making available the means of exploiting the important production potential of ASAL resources, thereby contributing significantly to income, employment and food security goals.
2. Reclaiming, where damaged, and protecting valuable and fragile natural environments.
3. Generating opportunities for improving the quality of life of present and future populations on a sustainable basis.
4. Determining ways and means of effecting symbiotic exchange of resources and products between ASALs and the high potential areas.

Specific issues related to pastoral systems development

The National Development Plan recognizes that livestock remains the most profitable way of utilizing the extensive rangelands of the

ASALs. Specific issues for pastoral development as outlined in the Sessional Paper No. 1 of 1986 include:

1. Disease control.
2. Rehabilitation of stock routes, sale yards, quarantine holding grounds, generation and better supply of market information and provision of weighing equipment to local councils to facilitate sale on a live weight basis.
3. Encouraging the private sector to establish commercial breeding and fattening projects with the specific aim of restocking after drought.
4. Encourage specialized business to identify and supply export markets with processed products. The Government is to encourage the establishment of dressed weight sales and chilling facilities at selected market centres.
5. Rehabilitation of livestock watering points and additional provision of such facilities intended to reduce range destruction caused by concentration of livestock numbers; in consultation with the pastoralists to examine possibilities for more extensive land use for livestock grazing in national parks and reserves or in buffer zones.
6. Promotion of bee keeping, small scale irrigation projects for supplementation of crop production, water harvesting techniques, silvi-pastoral and agroforestry systems.

Institutional framework

It is noteworthy that, because of the importance the government attaches to the development of ASALs, a new Ministry for Reclamation and Development of Arid and Semi-arid and Wastelands has been created. This ministry will assume the coordinating role of the Ministry of Planning and National Development, and also stimulate and participate in the implementation of ASAL Programmes in collaboration with various relevant technical departments (Odinga, 1989). The approach will follow closely on the District Focus for Rural Development strategy. As a departure from the initial approach, in which donor preferences took precedence over the peoples and governments priorities, the current approach is for the government to set up a special ASAL development fund into which donors or government funding will be channelled. It remains to be seen how these approaches will work. The development plan clearly specifies that, when this fund is finally established, districts will be expected to bid for their allocations under the fund for specified and integrated activities, reflecting the viability of proposals, the level of implementation and the effectiveness of previous allocations. This, it is hoped, will facilitate the generation of genuine beneficiary demand, thus ensuring their effective participation in identification, implementation and evaluation of the programmes/projects. In turn this will, hopefully, guarantee sustainability of proposed activities and projects. In addition the fund is expected to support a mobile extension training and management service team.

DEVELOPMENT MODELS: JUSTIFICATION, ACHIEVEMENTS, AND FAILURES

Until the late 1970s, when the government realized the need for a coordinated integrated development approach to address the often complex and intricate issues in ASALs, approaches to development in ASALs had mainly been sectoral. The revised approach has addressed issues such as infrastructure, human resource capacity, environmental conservation (mainly on cultivated land), water development, crop production, livestock production, rural enterprises, credit and extension services. The evolution of pastoral systems development models in Kenya, however, dates back to almost a quarter of a century ago, during the colonial period, and appears to have had a basis on Western ranch style systems. Basically, four models have been proposed and implemented, with varying degrees of success. The models are grazing schemes, group ranching, commercial ranching and grazing blocks. I examine these models in the paragraphs that follow.

Grazing schemes model

Grazing schemes were proposed and implemented by the colonial government in the ten year development plan (1945–55), when the colonial administration sensed deteriorating conditions in African reserves due to increasing population pressure. The schemes were established as models of proper land use and livestock management, in the hope that the pastoralists would be convinced of the necessity to reduce their livestock numbers. To overcome overuse of land resources, and to reduce soil erosion to manageable levels (Ole Sadera, 1986), pastoralists were organized, ethnically or traditionally, into large, organized areas with physical boundaries. Water and dipping facilities were provided. Each family in the scheme was alloted a livestock quota which they were not to exceed or face eviction from the scheme. Excess stock was to be sold. Members were to pay for purchase of acaricides, and also for salaries of borehole and dip attendants from proceeds of livestock sales.

Ole Sadera (1986) has summarized the success and failure of this model thus:

> Rotational grazing was not strictly enforced to induce the pastoralists to remain in the schemes. However, due to climatic conditions, the pastoralists continued to move out of the established schemes; the British administration did not have much control of the situation because the traditional chiefs were not prepared to part with the traditional way of life and, therefore, did not support the administration. Consequently, the schemes merely existed by names.

On the success of the Konza scheme in Kajiado district, he notes:

> Although the Masaai were suspicious at first of the objectives of the colonial administration, they appreciated the services provided in the scheme, particularly dipping and annual vaccinations of livestock because they reduced deaths due to tickborne disease. As a result, livestock numbers increased considerably beyond the capacity of the scheme. As would be expected, families with large numbers of livestock refused to sell the surplus and preferred to quit the scheme. Initially, four families quit, subsequently, movement out of the scheme became part and parcel of the management. The last six families left in the height of the 1960–81 drought and the scheme was subdivided in 1964.

Group ranching model

The group ranching concept was introduced after the failure of the grazing schemes. Although established along the same lines as the grazing scheme model, group ranches were established by an act of parliament (Group Representative Act, 1968), whereby land ownership was legalized and members could use land titles to obtain credit from financial institutions for ranch development and operation. Group ranches came to be defined as production enterprises in which a group of people jointly had freehold title to land, maintained agreed stocking levels, and tended their livestock collectively while continuing to own them individually (Ole Sadera, 1986). The objective of the group ranching model was to establish a system that would enhance modernization of livestock production, while still maintaining the traditional ways of life of the pastoralists. The group ranching model was financed under the Kenya Livestock Development Project Phase I (1968–1973) and Phase II (1974–1983) by the World Bank (Muriuki, 1988). Phase II was commissioned after the initial success of Phase I. The project provided facilities such as range water development, disease control and marketing services. Credit was also extended, for the first time, to the pastoralists to assist in the transition from subsistence to a market-oriented economy.

The success of the group ranching model was to be short-lived, for in 1978 during the second phase, the consequences of the severe drought began to emerge:

> Group ranchers moved in search of better pastures and water, abandoning what they called Agricultural Finance Corporation (AFC) steers—those purchased with loan money. Boreholes were left unattended. Engines and vital spare parts were stolen, pipelines were tampered with, and number of plunge dips were cracked. Other group ranches that were ready for financing became reluctant to take loans; consequently new development projects came to a standstill.

Outstanding loans were rescheduled and qualified personnel employed to manage the ranches following the recommendations of the World Bank review mission. Of the original target of 80 group ranches, only 27 had been financed by end of project.

A second unforeseen problem occurred much later as the human population in the group ranches increased and more second generation members had to be registered. Social frictions ensued among clan members according to age-group, clan, political leanings etc. Livestock numbers could not, therefore, be controlled. Management committees were manipulated by politicians and eventually the model broke down. In response to demands of the members, subdivision was seen as the ultimate answer. This exercise is still going on, irrespective of whether the new units are viable or not. Some of the pastoralists have started farming. The consequences of such a transition, although concievable, are yet to be determined.

Commercial ranching model

The commercial ranching model was implemented in Kenya at the turn of this century, following the enaction of laws that allowed for freehold and long-term leasehold grants for agricultural purposes. Most of these ranches were developed in the high potential rangeland of Kenya by immigrant populations after the alienation of the lands from the pastoralists. Most of the commercial ranches have been taken over by indigenous Kenyans since independence. Commercial ranches are categorized in three broad classes, namely: individual ranches, co-operative ranches and company ranches (Langat, 1986). An individual ranch is owned and operated by an individual who is the registered proprietor of the land and has given his herd a market-oriented structure. A co-operative ranch is made up of several families and derive their authority from the Ministry of Co-operative Development under the Co-operative Societies Act. Membership is by subscription of share capital.

Company ranches, on the other hand, are such that land is leased or bought from the government or a county council by a group of people who purchase shares in the company. Under company ranches three variants exist i.e. private company ranch in which membership is limited; public company ranch in which shares are open to the public; and directed agricultural company ranch, in which the number of shareholders is not limited but the company is limited by the total number of shares.

The commercial ranching model, despite its shortcomings, seems to be the only viable model for commercializing pastoral livestock production in Kenya. As Lloyd Timberlake observes in *Africa In Crisis*:

> Ranching seems only to have worked in Africa under a specific set of conditions; the ranches must be owned by private individuals managing for profit and assured of a market for beef and dairy products conditions which only exist in Kenya and Botswana.

I hasten to add that commercial ranching in Kenya was given financial impetus during both phases of the Livestock Development Project, to which I have already alluded. The main rationale was that these ranches would buy immature stock from northern and north eastern Kenya for fattening and slaughter, in what was known as the "flying herd operation". This was seen to be a contribution to destocking that was part of the grazing block development programme in the pure pastoral societies in northern Kenya, which we discuss next. It should, however, be noted that several factors have militated against this proposition, not least livestock quarantine procedures and other marketing technicalities.

Grazing block model

The grazing block model was proposed for ecological zones V and VI (Table 1), which are the main arid regions in Kenya and where the main form of land use is nomadic pastoralism. This concept was conceived in the early 1970s, almost at the same time as the group ranch model was proposed. The main criteria that determined whether a given rangeland qualified to be developed into grazing blocks was its potential class (Muriuki, 1986). Lands that could not be developed into better or more intensive ranches seemed to qualify for grazing block development. It is noteworthy that lands in these areas have not been adjudicated and officially registered for individual ownership. Grazing blocks were therefore restricted to the more arid districts of northern Kenya.

Table 1. *The area of each ASAL zone*

Agro-climatic zone	Per cent r/Eo (km)	Area	Per cent Country Area
IV	40–50	27,000	5
V	25–40	87,000	15
VI	15–25	126,000	22
VII	<15	266,000	46

r/Eo = Ratio of rainfall (r) to Evapotranspiration (Eo).
Source: MOA, 1989

The main objective of the grazing blocks was to increase livestock production through the provision of permanent and temporary water in a bid to achieve proper sustenance from the forage resources. Another aim was gradually to sedentarize the pastoralists, with minimal interference into their traditional way of life lest they resisted change. In this model blocks were delineated by physical boundaries. Each block on average covered 600,000 Ha. Graziers were to be registered in a given block based on traditional land rights, after other technical details had been worked out. The organizational concept was that the blocks were to be manned by a block management committee consisting of a block manager, district officer, elders and technical staff. Rotational grazing was to be practised by use of small grazing units where some areas were to be used as dry season grazing reserves. Water development programmes were to be undertaken by balancing available forage and water. Water pans and dams were to be developed in wet season grazing, while boreholes and large pans were to be developed in the dry season reserves. During the dry season boreholes were to be opened to allow for animals grazing within the reserves to water. As soon as vegetation was depleted the animals were to move to boreholes in another part to avoid localized overgrazing. During the wet season the boreholes were to be closed. Pastoralists benefitting from the facilities provided in the block were to pay a water use fee for maintenance of facilities provided, and were also to maintain a food livestock off-take by selling to Livestock Marketing Division (LMD). Veterinary services were also to be provided.

The initial benefits of the grazing block development were a reduction in mortality rates for livestock through disease control. Water provision reduced trekking distance of animals to water especially during dry periods and generally animal performance increased. Although technically viable, however, the block concept has not worked as intended for the following reasons:

1. Quarantine measures and security along trek routes affected sales, and therefore the anticipated off-take was not realized. Also LMD ceased to buy livestock.
2. Block development was not accomplished in all the stipulated areas. This uneven development resulted in an influx of nomads and their livestook to the more developed blocks, especially during drought. In some cases the geology of the areas was not properly studied, hence some dams never accumulated water. Therefore planned grazing patterns could not be enforced.
3. Government officials were unable to collect the water use fees required for maintenance of boreholes and dams. Dams got silted up, some pumps broke down and spare parts could not be bought. Grazing pressure increased on the few remaining watering points leading to overgrazing.
4. After donor assistance was terminated, logistic and financial problems militated against effective supervision of the blocks.
5. The pastoralists were not fully involved in the initial project planning. Traditional rangeland use and herding practices were not incorporated in project designs.

Grazing block development projects, although technically viable, lacked grassroot support from the pastoralists. They viewed themselves as beneficiaries rather than as participants. Hence, as soon as the donors pulled out, the envisaged operations of the project have not been sustainable. It is also important to note that block development was accomplished from 1970–1982 during the Livestock Development Project Phases I and II. The project terminated before covering, as had been intended, all the planned areas. To date, the future of the grazing block model as envisaged in Kenya is still uncertain. Donors seem unwilling to start from where UNDP/FAO stopped.

ALTERNATIVE APPROACHES

Although various development projects and models have been, and continue to be, implemented in the Arid Lands serious flaws in their design have meant that such "thought out" ideas by experts are not working. Environmental degradation and the threat of desert encroachment are permanent phenomena in some areas. Famine relief measures and rural poverty are the symptoms of a wider and complex problem that calls for alternative approaches for pastoral systems development. Recent attempts to diversify the pastoral economy through large-scale irrigation schemes and power generation projects, although important for the national economy, have often not involved thorough impact assessment studies in a wholesome context of interactions between the livestock owners and the would be "farmers". Such projects, although well meant, have often resulted in displacing the nomads from river valleys where they depend for dry season pastures for their livestock.

What, then, is the recipe for sustainable production, development and environmental conservation in ASALs? To answer this question, two realities should be born in mind. First, the trend of sedentarization of pastoralists, whether by design or default, is a reality. The second is the conversion of what is otherwise pastoral land to arable land, brought about by the desire to accommodate increasing population pressures and food production objectives in the shortrun. In my opinion, the future of the ASALs in Kenya lies in pastoralism (modified?) backed by proper environmental management, at least in the short and medium terms. The focus, I believe, should be on how to improve pastoral production systems in a way that will ensure sustainable production and human survival in the dry lands. The following approaches must, therefore, be part of an integrated development strategy for the pastoralists in the arid lands.

Indigenous knowledge systems

This involves the incorporation of local expertise and traditional risk management strategies in technology packages for pastoral development; positive traditional aspects of resource use and management should be identified and incorporated in resource management plans. Subjects such as "ethnobotany", "ethnozoology", "ethnohydrology" may be useful to consider, as they have a bearing on how pastoralists use resources and whether such use is reflected in the way resources are managed.

Community mobilization

This has to do with involving the pastoral community in identification of problem areas, project planning and implementation and continuity. Grassroot discussions with the people must have precedence. These should involve formal and informal education on project objectives and activities. Such mobilization can guarantee the sustainability and spontaneity of the development process. It is important to note that this approach is being addressed through the district focus for rural development. The approach, however, needs to be strengthened and widened in scope.

Appropriate projects/technology

These are specific projects and/or technologies that enhance survival of individual pastoralists. For example, key areas should include improvement of milk preservation and processing techniques to take care of the milk abundant during wet seasons. Water harvesting techniques, using rock catchments, and meat preservation fall into this category.

Alternative income sources

There is need to seek alternative income sources that can help to diversify the pastoral economy and reduce dependancy on livestock. Such diversification would enhance risk spreading especially during dry periods. For example, exploitation of drought adapted plants such as *Acacia senegal* and *Comiphora Spp* for gum production can supplement livestock keeping.

Drought management strategies

These should be deliberate strategies arrived at by the Central Government and other agencies aimed at mitigating the effects of drought, noting that scientists have not yet developed reliable tools for drought forecasting. These strategies should include hay preservation programmes, restocking programmes based on traditional loaning systems, timely livestock sale campaigns and the creation of drought reserves. It should be noted, however, that the drought management strategy concept is being addressed on a more regional scale in Eastern Africa by the Inter-Governmental Authority on Drought and Development (IGADD), an organization to which Kenya is a signatory.

Landscape planning

It is my opinion that a policy on landscape planning is most needed at this stage of pastoral development in Kenya. This will have to be done in consultation with the pastoralists. Such a policy would address such issues as viable settlement patterns, rules and regulations for ecosystem management based on customary law, water resources provision and watershed management. It would also help to harmonize development activities by non-governmental organizations (NGOs), that are partly to blame for declining environmental conditions which obtain in arid lands.

Research

Research in arid lands must be strengthened to provide information that can be used in the formulation of projects and extension services. Research input should be at every level of the development process.

To conclude, there is urgent need to re-evaluate the development approach to ASALs, especially as regards the future of pastoral communities. The attention given to pastoral systems development in the current National Development Plan appears to be a step in the right direction.

REFERENCES

Government of Kenya, 1986: *Economic Management for Renewed Growth.* Sessional paper No. 1. of 1986 Govt. Printer, Nairobi.

Government of Kenya, 1989: *The sixth National Development Plan 1989–1993.* Govt. Printer, Nairobi.

Hansen R. M., R. D. Child, and B. M. Woie (eds), 1986: *Range Development and Research in Kenya.* Proceedings of a Conference, Agricultural Resource Centre, Egerton College Njoro, Kenya, April 1–5, 1986.

Hjort A., 1989: "Life in Drylands: Environmentally Balanced Production in the ASAL". Paper presented at a *Workshop on ASAL Soil and Water Conservation Programme Development,* 18th–22nd September, Embu, Kenya.

Lamprey H. F., 1981: "The problem of livestock Night Enclosures in North-Eastern Africa". *Arid Lands Newsletter,* Office of Arid land studies. University of Arizona, Tucson.

Lamprey H. F and H. Yussuf, 1981: "Pastoralism and Desert Encroachment in Northern Kenya". *AMBIO* Vol. 10 No. 2–3.

Langat R. K., 1986: "Commercial Ranches in Kenya". In R. M. Hansen, R. D. Child and B. M. Woie (eds). *Op. cit.*

Lloyd T., 1985: *Africa in crisis: The causes, the cures of environmental bankruptcy.* Earthscan Publishers. London.

Muriuki R. M. M., 1986: "Grazing Block Development in Kenya". In R. M. Hansen, R. D. Child and B. M. Woie (eds). *Op. cit.*

Mwangata J. H. N., 1986: "History of Range Management in Kenya". In R. M. Hansen, R. D. Child and B. M. Woie (eds). *Op. cit.*

National Environment Secretariat and Human Settlements Secretariat (NEHSS), 1984: "Endangered Resources for Development". *Strategy Conference for the Management and Protection of Kenya's plant communities: forests, woodlands, bushlands, savannahs, and aquatic communities.*

Odinga O., 1989: "Frame for ASAL planning and Evaluation". Paper presented at a *Workshop on ASAL Soil and Water Conservation Programme Development,* 18th–22nd September 1989, Embu, Kenya.

Ole Sadera, P. L. K., 1986: "Group Ranches in Kenya". In R. M. Hansen, R. D. Child and B. M. Woie (eds). *Op. cit.*

Pratt D.J., 1967: "Rangeland Development in Kenya". *Annals of Arid Zones* Vol. 8 (2):177–207.

"Any Time for Space?" Farming Systems in Northern Ethiopia

Odd Eirik Arnesen

> Where land utilization practices are firmly established and have become the basis of a country's economy, the adaption of a new utilization programme conforming to the limits imposed by the natural environment may well involve a social and political revolution (Jacks, 1939)

The deteriorating livelihood of the people in the Sahelian countries has been a common concern since the first large-scale relief operations started in the early seventies. Analysis of the crises in the Sahel has, to a great extent, shown that the drastic manifestation of environmental degradation cannot simply be seen as stress on marginal physical environments. The acknowledgement that the problem is not merely a physical one with a complementary technocratic solution has not been emphasized, and there is still a long way to go before the ideological underpinnings of common concepts like land use management are acknowledged and that; "an understanding of nature is fundamental to the manner in which science sees and conducts itself, yet the concept is generally taken for granted" (Smith *et al.*,1980).

LAND USE MANAGMENT AND DEVELOPMENT

The complex problems facing the people of the Sahel have led to a call for holism and interdisciplinarity in the study of changing rural societies. It is clear that new approaches to environmental degradation are needed, as there are considerable problems in combining the study of essential physical and social processes which derive from the different epistemological and ideological characteristics of the natural and social sciences (Blaikie, 1981). One of the main purposes of this paper is to discuss the need to be able to foresee the researcher's position and the ideological and practical consequences of research. If this is not done I think that the plea to integrate nature and man in one holistic theory will not only be hollow but also threaten the local population, especially the poorest section of it. As such this is also an inherent critique of positivism as it has "re-entered" social sciences through the man/nature paradigm; or was it never really dead?

It is certainly no coincidence that the marginal areas coincide with politically marginalized and powerless people who have no realistic alternative to what they are doing within the existing political power structure. So the change of paradigm has to have further implications than just a shift in the researcher's focal point. "The technocratic approach" to environmental degradation will maintain political inequality and, at the same time as we continue with our "neutral" land use planning, shift the blame for its own failure onto its victims. This will create "illusions of concern on the part of those really to blame", allowing them to tighten the control over the marginalized poor in the name of "environmental protection". New "catch words" like "sustainability" seem to be in danger of being defined in the same technocratic manner although, because often presented as just more or less empty phrases, their ideological consequences are rarely identified (Baker, 1984).

Despite the radical analyses of Baker and Blaikie, which argue that environmental concern is often class biased so environmental degradation will continue unless it seriously threatens the accumulation possibilities of the dominant classes, we are still faced with the problem of how to relate "socialized" man to nature. While the modernization school and the radical dependency (Marxist) schools were irreconcilably antagonistic to each other on the meaning and mode of development, they seem to share the common view that nature is merely a resource for development (Shanmugaranam, 1989). The search for common ground has led to the rejection of both (state) socialist and capitalist development strategies by the eco-socialists or eco-anarchists who are looking for a type of ecotopia. Researchers, like Blaikie and Krokfors, have used theories labelled "political economy of soil erosion" or "political ecology" to analyse this problem of integration (1985; 1989).

In one sense this brings us to a third approach, based on references to "general systems theory", mostly used by natural scientists to enter the realms of social theory. Although, as a reaction to positivism and functionalism, rejected by most social scientists during the radical revolution this has been the first step outside the research station for the natural sciences. The quantitative nature of the models used has made demography an important variable, often used as the engine for change. Another tradition, basing itself on human ecology, came from within the social sciences using general systems theory to provide a "holistic view of Nature" (Watts, 1982). Based on an organic analogy these theories reflect a view of society as a living system and of man as an ecological dominant. Most of these researchers have adopted "a cybernetic view of adaptive process in social life" (Watts, 1982). Societies are seen as general purpose systems whose goal is nothing more than survival. This is clearly so in the reasoning of Kjekshus writing on East Africa (1977a). Populist oriented theories used by the "new-environmentalists" base themselves on one or more of the following points:

- an undifferentiated peasantry
- availability and superiority of indigenous knowledge
- peasant form of production as the basis for democratic development

In Ethiopia famines can be traced back to the ninth century (Pankhurst, 1972). Although Ethiopia had a different social structure to the rest of Africa, the argument made by Kjekshus cannot be said to fit. Societies which are in control of their environment will, on the other hand, not change, as argued in Boserup's theory of agrarian change, in which ecological stress, and even soil erosion, are needed to change to a more intensive form of production (Boserup, 1965; 1976).

SPACE AND TIME IN DEVELOPMENT

Next I briefly outline a redefinition of the mode of production theory, once so fashionable, by bringing in elements from Hagerstrand's "time geographical school" (1972). I do not wish to elaborate on the mode of production controversy here but just to point to the two basic assumptions used in this paper (See Klein, 1985). First, it is necessary to see history as a dialectical process in which change flows largely from the contradiction and internal tension within any given situation. Second, these contradictions are not always economic. As productive forces develop, the linkage between the forces and relations of production (which at any given moment are shaped by a variety of technological, ecological, spacial and historical factors) remains vital.

Any analysis of environmental degradation which deals with both man and nature has to include a "place-based" and a "non-place-based" analysis of the relations of production under which land is used.

> Geographers have for long claimed an interest in the differences which exist between places and there is an increasing awareness among them that many observable variations of phenomena in space are attributable to relative locations rather than the intrinsic qualities of the individual places. (Chisholm, 1968)

Following Chisholms distinction between absolute and relative space, this paper emphasizes that space and the political organization of space express social relationships but also react back upon them (Harvey, 1974). This Marxist based conceptual construction of space raises the question of whether the organization of space is a separate structure with its own laws of inner transformation and construction, or the expression of a set of relations found in some broader structure, such as relations of production (Soja, 1985). Space must be seen as embedded in the mode of production as a dialectical whole and thereby a part of the transitional process of the articulated modes of production. In the transition to a new mode of production a spatial reorganization of the production units and fields of operation is necessary to make the territorial structure and the new production relations compatible. Regional aggregation of people and resources has been the usual way of accounting for man/land ratios and estimating carrying capacities. But the aggregation of population, land and other resources for an entire region is a statistical abstraction. Neither the population nor the resources people utilize are concentrated in one place

> Different units of action in society (individuals, groups and organizations) are thus placed in an environment where resources are accessible only in certain spatial and temporal situations. General availability in the purely quantitative sense for an overall region is not enough. (Carlstein, 1980)

This comment is important for practical action. Any use of resources and people in human projects entails arranging activities in space and time, so that the right inputs are combined at the right time (synchronization) and places (synchorization). Thus it is important to develop an analytical method which can cope with resource allocation and utilization in a spatio-temporal matrix so making it possible to unveil the problems of the articulated spatiality of modes of production in the development process. Central to this theoretical development is to see the chronic tension between "fixity and motion" within the structured coherence of spatial configurations (Harvey, 1982).

Mabogunje (1980) notes that, "clearly, certain types of rural spatial structures, more than others, must make it easier to launch a country into an era of high agricultural productivity", which indicates that a spatial restructuring of the society will enhance the development process and be an "engine" in the transitional process. Similar thoughts can be found in a speech made by President Mengistu Haile Mariam:

> If we consider the history and trends of economic development throughout the world, it is only when people live close together, be it cities or villages, that they can pool their manpower, their money, their inventiveness together in order to make the best use of the resources for them.

PEASANTS IN TRANSITION

The role of agriculture in economic transition, the "agrarian question", has been widely debated (Lenin, 1976; Byres, 1982). The drive to socialism can be interpreted as a consequence of the limitations of capitalist development under conditions of dependency, neocolonialism and the attempt by African leaders to find another way forward (Munslow, 1986; Hyden, 1980): a process that can be described as the creation of conditions for a transition to socialism, defined by socialist researchers as a "road of non-capitalist development" (Gyenge, 1976; Solodovnik *et al.*, 1975). But socialist development strategy has not shown itself to be much more successful than its capitalist counterparts. A central debate has been whether this failure is based on the dynamics of the international economic system or the internal conditions specific to each country. To come to grips with these questions one has to develop a model that can incorporate analyses from the level of the household all the way up to international capitalism.

The time geographic model provides a basic tool for getting closer to the resource utilization by human populations in their social and natural environments (see Figure 1). A main element in the model is that the human population is looked upon as a web of paths which flow through a set of time-space locations. This time-space region contains the social system and is the setting of everyday life (Hoppe and Langton, 1986). Since it is a time-consuming process to move around in space, the individual-paths are always tilted upward in relation to the geographical plane. The paths of individuals are not isolated. They come into contact with one another and are coupled into bundles. These webs or paths can be seen on different scale levels as the setting in time-space of an individual's daily activities; the whole lifetime of an individual; the lifetime of a particular project, or a complete socio-spatial transformation (Gregory and Urry, 1985). In the settlement system stations, such as buildings or fields, can be depicted as pillars in time-space. This is not only related to the physical bodies that have location and structure in space and time, many legal rules and other symbolic or normative elements in society are also manifested in time and space. One example is the control areas or domains, which are the physical manifestation of norms, laws and contracts regulating the access to different time-space locations. Hagerstrand (1970) has defined a domain as a time-space entity within which things and events are under the control of a given individual or group. Some smaller domains are protected only through immediate power or customs, e.g. a favourite chair or a place in a queue. Others, of varying size, have a strong legal status: the home, land property, the premises of a firm or institute, the township, country, state, and nation. Thus there exists a hierarchy of domains.

The individuals whose paths intersect at stations come together to participate in projects; i.e. "the entire sequence of simple or complex tasks necessary to the completion of any intention-inspired or goal-oriented behaviour" (Pred, 1977). The performance of projects requires the synchorization and synchronization of different people and resources at stations, which thus comprise not simply positions or buildings but institutions for the accomplishment of projects requiring the organized interaction of people and the consumption or transformation of material resources (Gregory, 1978; Gregory and Urry, 1985). Thus projects function as allocative mechanisms in time and space for human activity and material resources (Carlstein, 1980). Over the short term the stations are conceived as fixed "pillars" in space and time.

People simply move between these points, or through this grid of pipes, according to a "choreography of existence" which is set by the given, but unspecified, needs of different projects and the operating constraints (Pred, 1977). The explicit constraints are of two types:

i. those which are properties of the individual human beings; the "capability", "coupling" and "authority" constraints;
ii. the scarcity of the resources of time and space available to one individual constrains his or her behaviour in the short term. The prediction of fixed stations as the "pillars" which give form to the pattern of paths and webs are created in response to these sets of constraints.

But the "choreography of existence" cannot be understood unless the stations are considered as more than points on a plane, and "unless the social rules which constrain and permit behaviour of different kinds by different people in relation to the resources allocated at these points are fully and explicitly introduced into the analytical scheme" (Gregory and Urry, 1985).

A MATERIALIST REVISION

Since the Time Geographical School has been introduced attempts to integrate it into more structuralist oriented research have been made, but still there are several reservations.

First, it operates with a naive and defective conception of the human agent. Hagerstrand treats individuals as coming into being independently of the social settings which they confront in their day-to-day lives. Agents are regarded as purposive beings in the sense in which their activities are guided by projects which they pursue. Second, Hagerstrand's analysis tends to recapitulate the dualism of action and structure, albeit in rather novel form, that plays an important role in Giddens theory of structuration (1984). "Stations", "domains" etc., are themselves taken as given, the outcome of uninterpreted processes of institutional formation and change. It places little emphasis on the essentially transformational character of all human action. Third, the concentration upon the properties of the body and its movement through time-space only in terms of constraints is unwarranted. All types of constraint are also types of opportunity, media for the enablement of action. This can be seen as a version of the historical materialist notion, that allocation of scare resources of the body and its media have some sort of determining effect upon the organization of social institutions in all types of society (a logic that closely resembles the villageisation policies of Ethiopia). Finally, time-geography involves only a weakly-developed theory of power. The notion of "authority constraints" linked to capacity invokes a zero-sum conception of power as a source of limitation upon action.

In this section I attempt to link time geography to the production process as part of the mode of production. Each mode of production attempts to form its own territorial structure with new forms of territorial integration. (Buch-Hanson *et al.*, 1977; Schmidt-Renner, 1977). But, at the same time, the territorial structure can have a restricting effect on the development of the new mode of production as the "inherited" and "projected space" (Lipietz, 1980). Production in a given locality causes physical and functional structures to be formed around it, including humanistic perspectives such as attachment to regions, social control or religious attachments. The physical and functional spatial structures, composed of the geographical localities and infrastructure, are the elements that constitute territorial structure (Buch-Hanson *et al.*, 1977). As a social product space can be continously reinforced or reproduced over time, and is thus both the medium and outcome of social action and relationships (Soja, 1985). Each mode of production thus is a result of past, present and future modes of production.

Given these conditions one can "re-interpret" the classical debate on the articulation of a peasant mode of production with other forms of production. Turning the coin, one can also see new spatial structures as means to develop new forms and modes of production.

AGRARIAN RESTRUCTURING IN NORTHERN ETHIOPIA

So far, I have focused on the theoretical aspects of changing man/land relationships. In this section I give some examples from the Ethiopian situation in general and from my own study area, Merhabete, as a case study.

One of the first objectives of the Revolutionary Government after 1974 was to integrate the peasantry into the Revolution (Griffin *et al.*, 1985). The government recognized the impossibility of constructing an egalitarian socialist society based on the rural structure it inherited. First priority was given to alter radically the production relations between the land and the tiller, a major slogan during the first stage of the Revolution (see Cohen *et al.*, 1976 for a different view of this issue). The Nationalization of Rural Land Proclamation of 1975 provided for the establishment of Peasant Associations (PA) throughout the country in which the peasants were to have the power. In most places this programme was welcomed and carried through by the peasant population under guidance of *Zemetsa* students, even though, as some researchers point out, the Revolution was initiated as a revolution from above (Ottaway, 1976). Initially, the reform made the peasantry more independent of other classes and thereby consolidated what many researchers identify as a distinct "peasant mode of production". Since then the Government has built new structures, trying to "capture" the peasants into national plans for development. On average it is estimated that no more than ten per cent of what each peasant household produces finds its way to the market, which indicates the level of economic integration! (Fassil, 1984).

Since the first years of the Revolution the government has encouraged the peasants to engage in collective forms of production. In the 1980s an increasing effort was put into this programme, together with a villageisation programme. This is a "second stage" in agrarian reform, integrating the "peasant mode of production" into the larger regional and national economy and breaking the peasant's "subsistence economy". The policy makes it clear that the development of agriculture rests on institutional development, particularly agricultural producers co-operatives. These are not only regarded as economic institutions, but also as "instruments of political, social and technological transformation of the peasantry" (Fassil, 1984). As Hyden put it: "In their case (the peasants) it is not only a matter of improvement of material conditions. It is also a question of losses in respect of other values, and above all it is a matter of trading social autonomy for increasing dependence on other social classes" (*ibid.*,1980:4). In many parts of Ethiopia this is reflected in peasants reluctance to join the co-operatives even when there are strong social and economic incentives to do so. On the other hand, different authors stress that land reform does not lead to increased production *per se* and agrarian reform, both in theory and practice, raises just as many problems as it solves, as has often been pointed out. Structural changes have not been a sufficient prerequisite for a socialist development strategy and the "Peasant mode of production" is able to survive and thrive. At the same time the present condition of famine and environmental degradation on a large scale, will force the government, not only to undertake structural changes, but to do so fast. The annual growth of population reached 2.9 per cent in 1985, and the highlands already support more than 78 per cent of the local population and 70 per cent of cattle, with the highest livestock density in Africa (FAO). These areas have a very unstable ecological farming system based on different levels of intensification of the original "ley" system of farming. More than two thirds of all Ethiopian farmland is on slopes of over 25 degrees. About half the highlands are already significantly eroded, and now reach unprecedented levels. It is estimated that 600,000 hectares of cropland disappear each year and, if this process is not reduced, about ten millon people in the highlands are going to be unable to feed themselves by the year 2010.

Plough based agriculture was the basis for the wealth of the old empire. But the livestock density is now a contributing factor to the increasing rate of erosion through overgrazing. Cattle ownership also encourages the increasing rate of economic differentiation that is found in the PAs; non-owners have to rent oxen from the wealthy members of the PA, which leads to "erosion" in the socialization process!

A PEASANT MODE OF PRODUCTION?

The peasant sector accounts for more than 90 per cent of the farm total, holdings are quite small, often less than one hectare. A PA is made up of 800 hectares of land, in some areas more, with between 200 and 500 households. Until 1977 the PAs acted relatively independent of national government bodies. They had, and still have, their own executive bodies, juridical bodies and security agencies. In 1977 the All-Ethiopian Peasant Association (AEPA) was founded to co-ordinate the activities of the many PAs. At the same time this increased the governments ability to control and integrate the leadership in the PAs, and to direct the mobilization of the peasantry based on centrally designed plans. The peasants do follow the directives given through the PA leadership to a certain extent but the leadership is certainly not a grass roots one.

The second form of production is the Producer Co-operative (PC). As few as three households may pool their resources to form a PC. There are different stages in development of co-operatives called *Malba, Welba* and *Weland*. The "First Stage Co-operatives", *Malba,* consist of collectively cultivated land with some land privately "owned". Implements are hired from the individual peasant by the co-operative. Peasants share the collectively generated income on the basis of their labour contribution. The "Second Stage", *Welba,* involves a transfer of the means of production to the co-operative so that they are owned collectively. The final stage is the *Weland* in which all land of several *Malba* and their means of production are brought under its control, and each member of the

Welba is a member of a *Habre* (working brigade) under the central *Weland*. Note that a *Malba* co-operative can operate within the limits of the PA. The co-operatives are allotted the best land through a redistributive process; one of the subsidies designed to attract the peasants to become members.

The concept of villageisation is by no means new in the African setting. A spatial reorganization of the production units has been a component of the policy of Agrarian Transformation in several African states. The first notions on villageisation in Ethiopia are found in the Land Reform Proclamation no 31 of 1975, in which the PAs were given the responsibility to undertake villageisation programmes. Initially this had little effect, as the main concern of the peasantry was directed towards redistribution of land and abolition of tenancy. As most of the peasants in the northern highlands simply retained the land they had before the revolution, there was only a minimal amount of redistribution. The Ethio-Somalia war made it necessary to move people into villages for strategic reasons in Bale and Hararge. No guidelines on villageisation were given. It was not until July 1985, after the third plenary session of the central committee of the WPE, that the villageisation programme was intensified for all regions (Berihun, 1988). The programmes up to this period were based on *ad hoc* local operations and security objectives. The increasing pace of the villageisation campaign made it necessary to coordinate, control and guide the programme. Thus the National Villageisation Central Committee (NVCC) was formed in 1986.

Up to January 8, 1988, 26.9 per cent of the rural population had been moved into new villages. Villageisation policies have become a major strategy for transformation of agriculture in Ethiopia:

> The obstacles to growth in the agricultural sector of the economy are disjointed land holdings, scattered living pattern, the absence of adequate trained manpower, technical backwardness and the dearth of basic development activities. One of the—indeed the major one—taken to tackle these problems is villageisation. (Fikre Selassie Wogderess, Chairman, National Village Coordinating Committee, 1987)

The objectives and rationale for the villageisation campaign are set forth in an Amharic document entitled *Villageisation Guidelines* (NVCC,1985 see Cohen *et al.*, 1987). There are four major objectives: (i) to enhance extension services aimed at increasing agricultural productivity, (ii) to promote more rational land use patterns and conserve natural resources, (iii) to facilitate access of rural people to schools, clinics, water supplies and service co-operatives, (iv) to strengthen security and self-defence. There is no direct linkage between villageisation and co-operatization in the documents related to villageisation but; "the settlement pattern is scattered, individualistic and lonesome and, consequently, the peasants lack a sense of belonging and social feeling". (*Ibid*:9)

According to the Ten Years Perspective Plan, 53 per cent of the total farming population will be members of a PC by 1993/94, this low performance is expected to be improved by villageisation. The pace of this planned process has, however, slowed down. According to the latest Three Year Development Plan (1986/87–1988/89, ONCCP 1987) producers are planned to increase their share in the total area cultivated and in the total crop production from 2.29 per cent and 2.24 per cent respectively, to 5.79 per cent and 6.75 per cent respectively: "the role that will be played by villageisation in the multi-faceted efforts to improve this poor performance and score great successes will be extremely great". (NVCC, 1986).

The villageisation programme in Ethiopia can also be seen in the context of a physical and regional planning procedure. Regional planning became a part of the overall planning system after 1984—but is still in its infancy (two regional plans in 1984 (Abese Teferi, 1987; Sileshi Tefera, 1987). The multipurpose trans-sectoral villageisation programme has thus been carried out without any linkage to any overall regional plans. Veljkovic (1987) also points out that:

> Planners and politicians should be equally aware that we in socialist countries are often prone to follow some political targeting notwithstanding the objective conditions and resources for carrying out those political tasks.

Both good solutions and mistakes in physical planning are long lasting, but these long term consequences might not appear for a long period.

In this context it is interesting to note what Kjekshus writes about the villageisation campaign in Tanzania and its possible consequences:

> Nucleated settlement will mean overcrowding of restricted areas with people and domestic animals and the accompanying soil erosion, gulley formations and dust bowls which are all common features in situations where human initiative has suddenly overtaxed the carrying capacity of land without compensatory inputs to increase the quality of cultivation. (Kjekshus,1977b)

There has not been any large-scale assessment of the programme except for the Arsi region. There the SIDA mission's overall conclusion was that the process of moving people into villages, in itself, had no short term measurable effect on overall production. The same is argued by Hyden for Tanzania (Hyden, 1980). Contrary to rumours and statements in the press, there was no evidence of villageisation being accompanied by collectivization.

One of the main elements in the critique of villageisation policies is the increasing distance this creates to each farmer's plot. A study from Selale (Shewa) points to more specific negative aspects of this reorganization summarized by Berihun, (1988) as follows. First, far away plots are put out of cultivation (often attacked by animals). Second, the practice of using dung on farm plots as manure has been abandoned. Third, distance to pasture increased on average eight-fold. Fourth, lack of common pasture adjacent to the village. Fifth, decline in livestock, 27.3 per cent sold to buy material for the new house, leading to increased use of fuelwood because of the shortage of dried dung, the main fuel source.

Farmers also reported crop losses due to problems of guarding the land. Another "non-productive" aspect was the increasing distance from water sources, so the consumption of water dropped. An increasing prevalence of diseases might be the result of such a development as is clearly recognized by the peasants.

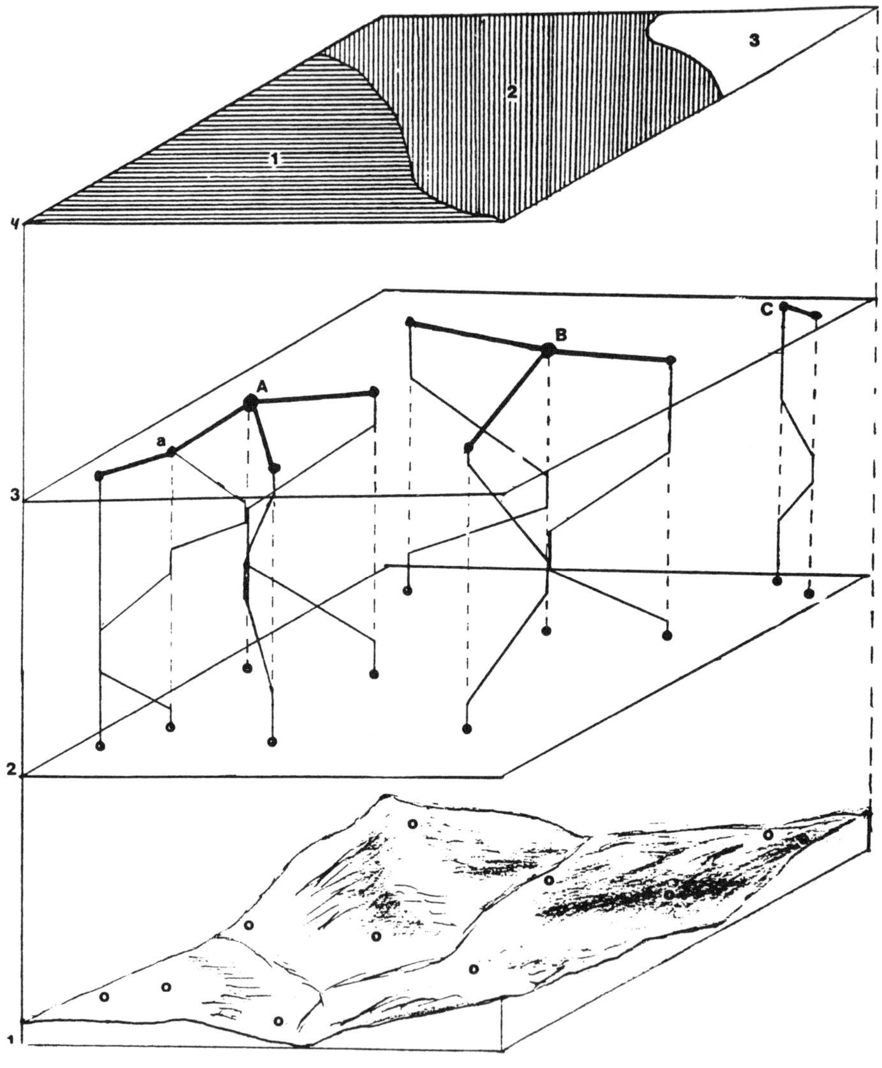

Figure 1. *Levels of Abstraction in a Geographical Setting*

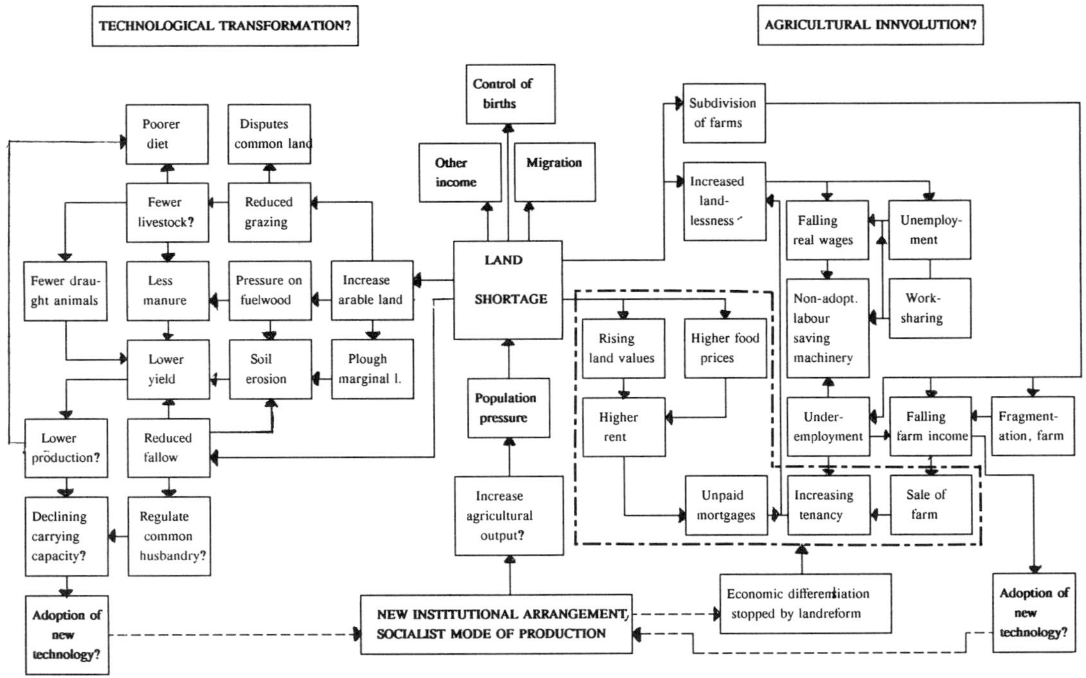

Figure 2. *A model for agricultural transformation and involution in Ethiopia* (based on Grigg, 1980)

As seen on the left side in Figure 2, the differentiation process partly stopped by the land reform will then become effective and will have implications for the left side of the figure as access to animals and grazing land will be based on new economic and social power structures. Any policy option now emerging must take into account that the peasant society is not homogenous nor is it static, and any change in society will have implications for the territorial structure of the society at all levels.

MERHABETE—A CASE STUDY OF AGRARIAN RESTRUCTURING

I will now try to use my own study in Merhabete to sum up some of the main points that have been raised in this paper. Merhabete is located in the Central Highlands of Ethiopia and, in the way that it is dissected by river gorges sloping westwards into the Blue Nile, is typical of the area. It is also typical in that most of the population was dependent on food aid in the period 1985–1989; although trucks going to Merhabete with food aid met government trucks bringing compulsory grain quotas out of the area!

The population is partly Amhara and partly Oromo. Due to malaria in the lowlands, most of the settlements are on the plateaus. As the map shows, most of the PAs have a boundary stretching as far as possible from the highland down to the rivers, forming a catena. The Land Reform redistributed the land so that each household was entitled to six different plots, ranging from 2,600 m to 1,400 m above sea level, in addition to their home garden. This gives the peasants a multiple choice as to which type of crop to grow and where. As there are two cropping seasons (*Belg/Kriemt*) in some years, this gives the peasant relatively good protection so he can at least expect some grain from his land. Redistribution of land has stopped due to land shortage, and new families can only obtain land from their families of origin or if some people die in the PA without heirs. As such people only get one or two parcels to cultivate, this clearly makes them more vulnerable to a crop failure than the older established peasants in the area.

The same logic can be found in most parts of the area, except for PAs on the tops of the mountains. There the households will have fewer parcels but will rely on specialization of crops to exchange with lowland farmers. This highland/lowland interaction also extends to arranging marriages and herding cattle in the lowlands. The reverse problem one can find in PCs that have got a consolidated holding near the village, in which members of the PC try to get fragmented land situated in different zones of the catena. In the lowlands in the west there is a totally different farming system, based on "slash and burn farming", made "rotational" with between 3 and 13 year fallow periods.

There is an acute shortage of both fuelwood and grazing areas in the highlands. This has had two effects: (i) animals are kept in the lowland for longer or shorter periods, often seasonally (marked by stars on the map) and often guarded by specialized herdsmen. Night sheds are moved around to get the best effects from the cow dung, (ii) very little cow dung is available in the highlands and what there is, is used for cooking. Further to the east, in Menz, this has led to the creation of a market for cowdung as fuel (Pankhurst, 1989).

Villageisation has mainly taken place in the highland areas and does not differ much from other large-scale settlements. In some areas villageisation has mainly been accomplished by moving existing houses into line. Due to the terrain and the high degree of fragmentation villageisation will not have the same impact on distances to the fields in Merhabete as it has had in other areas of Ethiopia. A study of a traditional village in Northern Shewa (Andit-tid area) showed a positive correlation between distance and production but, as it did not include the *Guaro* (the homestead garden), the results cannot be said to be conclusive (Yohannes, 1988). It does however indicate that distance is not only perceived in a quantitive way, to which I will return.

It seems that in Merhabete very little change has taken place since pre-revolutionary times in the actual land use pattern found on the map. The new settlement pattern now being introduced will mainly have short term effects on the ecology of the area. If the promised new infrastructures for the new settlements do not arrive soon, some problems might arise as to health. So far villageisation has not increased the peasants willingness to join the PCs. The reason, that the individual peasant is more productive than the PC, both in Merhabete and the county as a whole, is clear. The low performance of the PC might be related to the fact that it is the least productive farmers that tend to join the PCs; although I found other cases where rich farmers use this as an opportunity to make financial gains by obtaining irrigated land.

So the statements made both by Kjekshus and the chairman of the NVCC seem to be wrong in the case of Merhabete. No large-scale disruption of the local ecology or increase in production, by adopting new technology or people joining the PC, has taken place. I will, however, refer to a student who told me that he was impressed the first time he saw a villagized village, as the houses stood like soldiers ready to move out to the field together. The image of discipline as a result of a new environment was also found in PCs in a study undertaken by Eva Poluha, where the introduction of regulated work hours brought about a change in people's conceptions of both time and work in the context of the agricultural calendar. The peasants are transformed into labourers paid according to the time worked and are not related to any parcels of land with which they could identify. The same can be said about villageisation, individual households used to live in clusters of between 2 and 30; households related to the patron–client relationships found in cultivation activities and ritual activities (Adbar spirits) based on the territory of the village. For the poorer households to seek contact on a "class" basis outside the *mender* (=village) would be difficult.

Can we conclude that the changes that have taken place as part of the government's development efforts have failed, and that a peasant mode of production is still functioning in the area? If one looks briefly on the social side there are clear signs of processes often described as "Agricultural Involution", the term used by Geertz (1963) when more and more people care for the same resources and, to some extent, are able to increase production. What Geertz does not include in his analysis is that this process also includes social differentiation.

As regards the agrarian restructuring of the Ethiopian farming system, the present government has admitted the policy failures made so far. The attempt to change society along a socialist path by the creation of new spatial structures has not been successful. The new "socialist" territorial structure could not transform the territorial structure of the "peasant mode of production" and "force/guide" the peasants into a new way of production. Whether this is a result of the "peasant mode of production", through its passive resistance to new governmental structures, or the weakness of the Ethiopian state, will be a matter for debate in the years to come.

The structures that emerged in the rural areas as part of the transnational policies are now falling apart as the government is loosening its grip on the administrative structure. It is uncertain what will be left of the old power structure and what will develop out of it. Based on findings from my study area, a sort of agricultural involution, involving a process of differentiation, based mainly on the access to capital (mainly oxen) has taken place both within the nuclear and the extended family, but has partly been checked by the usufruct rights guaranteed through the PA. If the privatization process now emerging is not controlled, the process will gain speed and increase the distance between those who have and those who have not been in the PAs. The social differentiation is mainly related to the use of oxen and the limited access there is to livestock in periods of stress. So the peasant economy is facing a crisis made by its own "logic". If this crisis can be solved by increasing intensification of the cropping system by terracing (40–50 per cent increase in yield), or if this has to involve a total reorganization of the mode of production remains to be seen. The result will be decisive for the whole nation and the March speech made by President Mengistu, which announced a loosening of the role of PCs in the development of the rural areas and allowed private ownership of land, might induce rapid change in the rural areas of Ethiopia.

CONCLUSIONS

The attempts made to see man/nature in one holistic theoretical tradition in the 1980s have mainly been made from the side of the natural sciences. The new academic institutions emerging as the environmental crisis develops, both in the industrial and the developing world, have used "General Systems Theory" as an important tool to get closer to the problem of integrating natural and social sciences. While social sciences are involved in a subject–subject relation, "the theories and findings of the natural sciences are separated from the universe of objects and events which they concern" (Giddens, 1984). The result is that the relationship between scientific knowledge and the object world will become a technological one. This is clearly seen in the prestige of the "management" tradition within these types of institutions, in relation to governments and the problematics of social sciences "subject–subject relationship". The apparently neutral principles of management make it unproblematic to put the blame for its own failure on the undifferentiated subjects.

I have tried to show that any theory trying to analyse Man and his relation to environment will have to analyse the properties of the absolute location, the relative location according to human activities and the relation between the location and the modes of production as the society re-shapes its geography.

Table 1. *Pace of villageisation 1986–1988* (resettlement schemes not included)

	June 1986	June 1987	January 1988
No. of villages constructed	9,438	15,919	17,953
No. of villageized families	1,009,975	1,745,385	1,879,953
No. of villageized population	5,454,111	8,345,128	8,885,030

Source: Berihun, 1988.

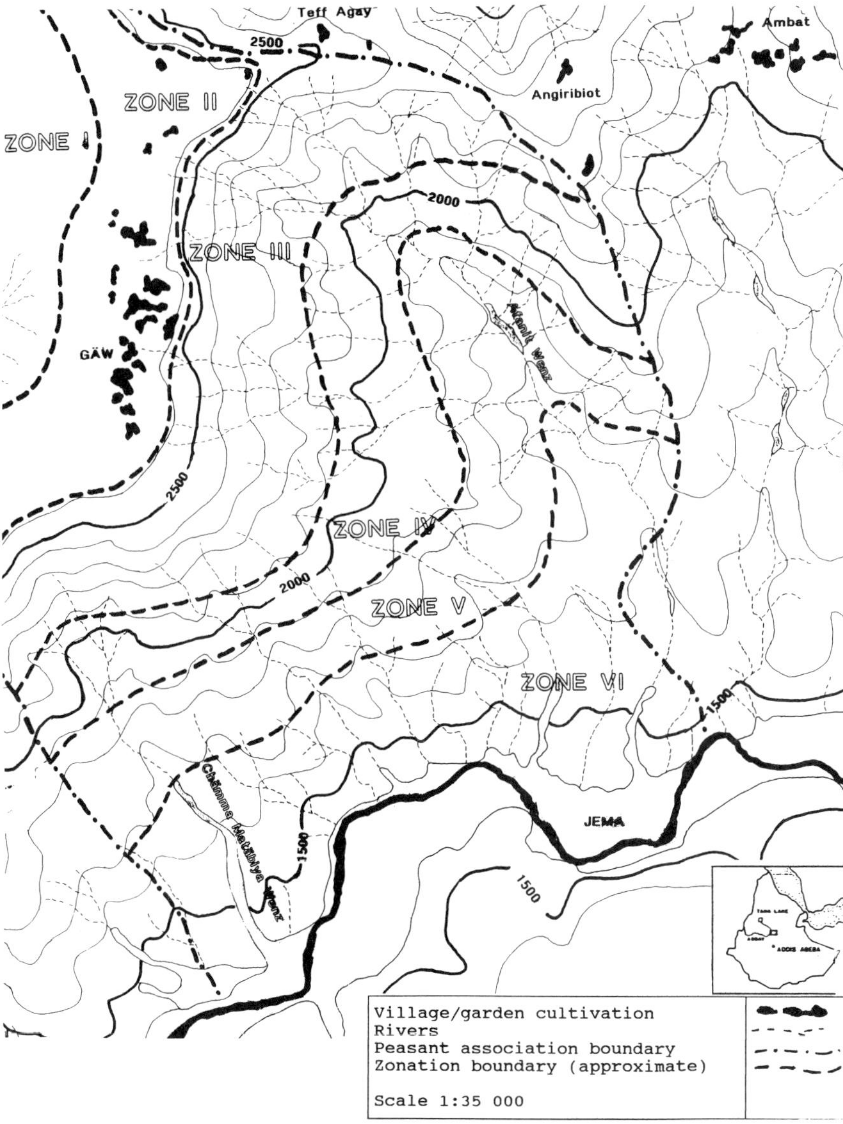

Map 1. *Division of land according to inheritance within six different zones in Gaw PeasantAssociation, Merhabete Awraja, Shoa*

REFERENCES

Abese Tefera, 1987: "Lessons from past co-ordination of regional plans and possible prospects", *ONCCP*:67–99.

Baker R., 1984: "Protecting the Environment Against the Poor, Historical Roots to the Soil Erosion Orthodoxy in the Third World", *The Ecologist* 14, 2:53–60.

Berihun Teferra, 1988: *An Economic Study of the Villageisation Programme in Ethiopia. The Case of a Village in Selale Awraja of Shoa Region*, B.A. thesis, Addis Ababa University.

Blaikie P, 1981: "Class, Land Use and Soil Erosion", *ODI Review* 2:57–77.

Blaikie P, 1985: *The Political Economy of Soil Erosion in Developing Countries*, London. Longman.

Boserup E., 1965: *The Conditions of Agricultural Growth*, George Allen and Unwin. London.

Boserup E., 1976: "Environment, Population and Technology in Primitive Societies", *Population and Development Review* 2,l.

Buch-Hanson M., B. Nielsen, 1977: "Marxist Geography and the Concept of Territorial Structure", *Antipode*, 9,2.

Burgess R,.1978: "The Concept of Nature in Geography and Marxism", *Antipode* 10,2:1–11.

Byres T. J., 1982: "Agrarian Transition and the Agrarian Question", in Harris, J., *Rural Development: Theories of Peasant Economy and Agrarian Change*, London. Hutchinson.

Carlstein T., 1980: *Time Resources, Society and Ecology*, George Allen and Unwin. London.

Chisholm M., 1968: *Rural Settlement and Land Use*, London.

Cohen J. M., A. Goldsmith, J. W. Mellor, 1976: "Revolution and Land Reform in Ethiopia: Peasant Associations, Local Government and Rural Development", *Rural Development Occasional Papers* No. 6, New York.

Cohen J. M., N. Isaksson, 1987: *Villageisation in the Arsi Region of Ethiopia*, Report prepared by SIDA Consultants to the Ethio-Swedish Mission on Villageisation in Arsi Region, December, 1986. Rural Development Studies no. 19, Swedish University of Agricultural Sciences.

Dessalegn Rahmato, 1984: *Agrarian Reform in Ethiopia*, the Scandinavian Institute of African Studies, Uppsala & Red Sea Press. New Jersey.

FAO, 1984: "Ethiopian Highland Reclamation Study. The Degradation of Resources and Evaluation of Actions to Combat it". Working Paper 19, by M. Constable.

Fassil G. K., 1984: "Mobilizing the Peasantry for Rural Development: the Ethiopian Experience in Progress", *Proceedings of the 7th International Conference of Ethiopian Studies*, Uppsala, Addis Ababa and Michigan. Ed. Sven Rubenson: 577–586.

Geertz C., 1963: *Agricultural Involution. The Processes of Ecological Change in Indonesia*, University of California Press, Berkeley and London.

Giddens A.,1984: *The Constitution of Society. Outline of the Theory of Structuration,* Polity Press. Cambridge.

Gregory D., 1978: *Ideology, Science and Human Geography,* Hutchinson. London.

Gregory D. and J. Urry, (eds), 1985: *Social Relations and Spatial Structures,* MacMillan. London.

Griffin K. and R. Hay, 1985: "Problems of Agricultural Development in Socialist Ethiopia. An overview and suggested strategy", *Journal of Peasant Studies,* 13, 1.

Grigg D., 1980: "Migration and Overpopulation", in White, P. and Woods, R., *The Geographical Impact of Migration,* London.

Gyenge Z., 1976: *Ethiopia on the Road of Non-Capitalist Development,* Studies on developing countries no. 90, Institute for World Economics of the Hungarian Academy of Sciences, Budapest.

Hagerstrand T., 1972: *The Impact of Social Organization and Environment Upon the Time Use of Individuals,* Plan International, Stockholm.

Harvey D., 1974: *Social Justice and the City.* Edward Arnold. London.

Harvey D., 1982: *The Limits to Capital,* Blackwell, Oxford.

Hoppe G. and J. Langton, 1986: "Time, Geography and Economic Development: The Changing Structure of Livelihood Positions on Arable Farms in Nineteenth Century Sweden", *Geografiska Annaler* 68, 2.

Hyden G., 1980: *Beyond Ujamaa in Tanzania,* University of California Press, Berkeley and Los Angeles.

Jacks G.V., 1939: *Vanishing Lands. A World Survey.*

Kirsch O.F., F.V. Goricke, J.F.G. Worz, 1989: "Agricultural Revolution and Peasant Emancipation in Ethiopia", *Studies in Applied Economics and Rural Institutions,* 21, Saarbrücken.

Klein M. A., 1985: "The Use of Mode of Production in Historical Analysis", *Canadian Journal of African Studies,* 19 (1), 9–12.

Klein M. A., 1980: *Peasants in Africa. Historical and Contemporary Perspectives,* Sage. London.

Krokfors C.,1989: "Population and Land Degradation", *Ecology and Politics. Environmental Stress and Security in Africa,* 197–210, eds. Hjort, A. and Mohamed Salih, M.A., Uppsala.

Kjekshus H., 1977a: *Ecology Control and Economic Development in East African History,* Heinemann. London.

Kjekshus H., 1977b: "The Tanzanian Villageisation Policy. Implementational Lessons and Ecological Dimensions", *Canadian Journal of African Studies,* XI, (2):269–283.

Lenin V. I., 1976: *The Agrarian Question and the Critics of Marx,* Moscow.

Lipietz A., 1980 (1975): "The Structuration of Space. The Problem of Land and Spatial Policy", in Carney, Hudson and Lewis (eds), *Regions in Crisis.*

Mabogunje A.L., 1980: *The Development Process. A Spatial Perspective,* Hutchinson. London.

Munslow B., 1986: *Africa: Problems in the Transition to Socialism,* Zed Press. London and New Jersey.

ONCCP (Office of National Committee for Central Planning), 1987: Workshop on Regional Planning in Ethiopia, Nazareth, (Unpublished papers).

Ottaway 1976: "Social Classes and the Corporate Interest in the Ethiopian Revolution", *The Journal of Modern African Studies*, 14, (3).

Pankhurst R., 1972: "The History of Famine and Pestilence in Ethiopia Prior to the Founding of Gonder" *Journal of Ethiopian Studies*, X, (2), Addis Ababa.

Pankhurst H., 1989: "What Change and for Whom? The case of women in Menz", paper presented at the conference on Options in Rural Development in Ethiopia, Bergen.

Poluha Eva, 1989 "Central Planning and Local Reality. The Case of Producers Cooperative in Ethiopia", *Stockholm Studies in Social Anthropology*, 23.

Pred A.,1977: "The Choreography of Existence: Comments on Hagerstrand's Time-Geography", *Economic Geography*, 53:207–21.

Schmidt-Renner G., 1977: *Om den samfunnsmessige lokalisering*, Aarhus.

Shanmugaratnam N., 1989: "Sustainable Development—A Conceptual Overview", *Proceedings of the Symposium on the Sustainability of Agricultural Production Systems in Sub-Saharan Africa*, 25–43, Norwegian Center for International Agricultural Development, Ås.

Sileshi Tefera, 1987: "Regional Planning and Regional Development Policies in Ethiopia: History, Problems and Prospects", *ONCCP*, 21–51.

Smith N. and P. O'Keefe, 1980: "Geography, Marx and the Concept of Nature", *Antipode*, 12,.2, 1980:30–39.

Soja E. W., 1985: "Spatiality of Social Life. Towards a Transformative Retheorisation", in Gregory and Urry (eds): 90–128.

Solodovnikov V. and V. Bogoslovsky, 1975: *Non-Capitalist Development: A Historical Outline*, Progress Publishers. Moscow.

Veljkovic J., 1987: "The Role and Scope of Physical Planning in Ethiopian Conditions", *ONCCP* :67–99.

Watts M., 1983: "'Good Try, Mr Paul'. Populism and the Politics of African Land Use", *African Studies Review*, 26, (2):73–83.

Yohannes Gebre M., 1989: *Land Use, Agricultural Production and Soil Conservation Methods in the Andit Tid Area, Shewa Region*, Soil Conservation Research Project, Research Report 17, Bern.

Ecological Implications for Tana River Basin Forestry and Irrigated Agriculture

Stig Johansson

INTRODUCTION

This paper deals with some of the ecological and social implications of the development of the Tana River Basin in eastern Kenya, with special emphasis on forestry at the Bura Irrigation Settlement Project. Specifically, it discusses some of the causes and effects for environmental degradation which are often associated with irrigated agriculture in arid and semi-arid areas, including the effects of other catchment development efforts. The effects of this degradation on the management of the irrigation scheme, the tenant farmers as well as on the surrounding pastoralist and traditional agricultural communities, are also discussed.

The need to support growing rural populations which are dependent on agriculture has increased the pressure to develop marginal areas in Africa by irrigation. Moreover, irrigation has been seen as a means towards increased food self-sufficiency as well as a source of foreign exchange earnings from cash crops. In Africa, only about 10 million ha, or 5.6 per cent of the agricultural lands are irrigated, as compared to 30 per cent in Asia and 19.7 per cent in developing countries as a whole (FAO, 1985). In sub-Saharan Africa, only an estimated 5 million ha are under irrigation, which represents about 15 per cent of the technical irrigation potential (FAO, 1986b).

In Kenya, the irrigation potential is estimated at 350,000 ha, of which 49,000 ha, or 14 per cent of the potential, is currently utilized. The modern schemes account for about 21,000 ha, whereas the small–scale or traditional schemes account for 28,000 ha (FAO, 1986b). Production in the major schemes (11,000 ha) include rice (Mwea, Ahero/Wescano, Bunyala), cotton and maize (Bura, Hola), and onion and maize (Perkerra). The remaining 38,000 ha produce vegetables, fruits, coffee, sugarcane or act as experimental schemes.

The environmental impact, with subsequent socio-economic repercussions for the development of irrigated agriculture, is attracting increased interest (Adams and Hughes, 1986; Hughes, 1990). Forestry, including irrigated forestry, as well as the management of the woody resources in connection with irrigated agriculture is also receiving increased interest (Armitage, 1985; Hughes, 1987). The first records of major irrigated forestry plantations are from the Indus Valley in the late 19th century. In Africa, the oldest plantations were developed at the Gezira Scheme in the Sudan in the 1930s. Currently they cover an estimated 0.43 per cent of the agricultural area (Armitage, 1985). The majority of the irrigation schemes are, however, still virtually devoid of trees. Despite a high demand for fuelwood as well as other forest or tree-related products (charcoal, construction wood, amenity trees), it has been estimated that the irrigated forestry plantations amount to less than one per cent of the area under irrigated agriculture (*ibid.*, 1985).

Since fuelwood demand is not satisfied within the agricultural schemes, the necessary supply comes from the surrounding areas. The result of this is that often vast areas of development induced desertification spread as concentric rings around the schemes. Additional problems are often land use conflicts between the traditional population, which makes its living from the natural resources, and the farmers who particularly need fuelwood.

In Kenya, the first trials with forestry in connection with irrigated agriculture were started at the Tana Irrigation Scheme in Hola in the mid–1960s. The work was neglected, and in the late 1970s no results or records of monitoring were available (MacDonald and Partners, 1977). Currently the only irrigated forestry plantations and research on forestry in irrigated agriculture are located at the Bura Irrigation Settlement Project.

ECOLOGY

The Tana River rises on Mount Kenya and the Aberdare Mountains and flows for about 650 km across the slopes of the highlands and the alluvial lowland plain into the Indian Ocean. It is Kenya's largest river with a total catchment area of 94,700 km^2, and the only one able to support large-scale irrigation development. An estimated two per cent of the basin is irrigable, primarily in the lower basin, and this represents approximately 50 per cent of Kenya's irrigation potential (TARDA, 1980; cf. Hughes, 1987). There is considerable variation in climate and topography along the course of the river. The lower Tana Basin, including the Bura area, is semi-arid to arid with bimodal annual rainfall of about 300–400 mm. Temperatures and evaporation rates are high, with little annual variation. The natural vegetation consists of a 1–2 km wide groundwater fed floodplain forest along the Tana and, away from this, mainly of *Acacia-Commiphora* bushland with low biomass production potential. See Map 1 below.

The riverine forest is a relic of forests which were once more widespread. It is the only high forest in the area and has considerable conservation value. Animal species, such as the Tana Mangabey (Cercocebus galeritus galeritus) and the Tana River Red Colobus (Colobus badius rufomitratus), are classified as "critically endangered" and "rare" respectively by the IUCN (1978). Plants such as the Tana River Poplar (*Populus ilicifolia*), which is endemic to the area, and one provenance of *Acacia tortilis* are also classified as endangered (FAO, 1986a).

POPULATION

The traditional population in the Bura Division consists of the Malakote and the Orma. There are occasional in-migrations of Somali pastoralists during the dry season. The population estimates for the area are rather unreliable.

The Malakote, who number about 15,000 in the scheme area (Hughes, 1985), are sedentary agriculturalists, practicing flood recession agriculture along the river levees and the ox-bow lakes, which are created from cut-off meanders during floods.

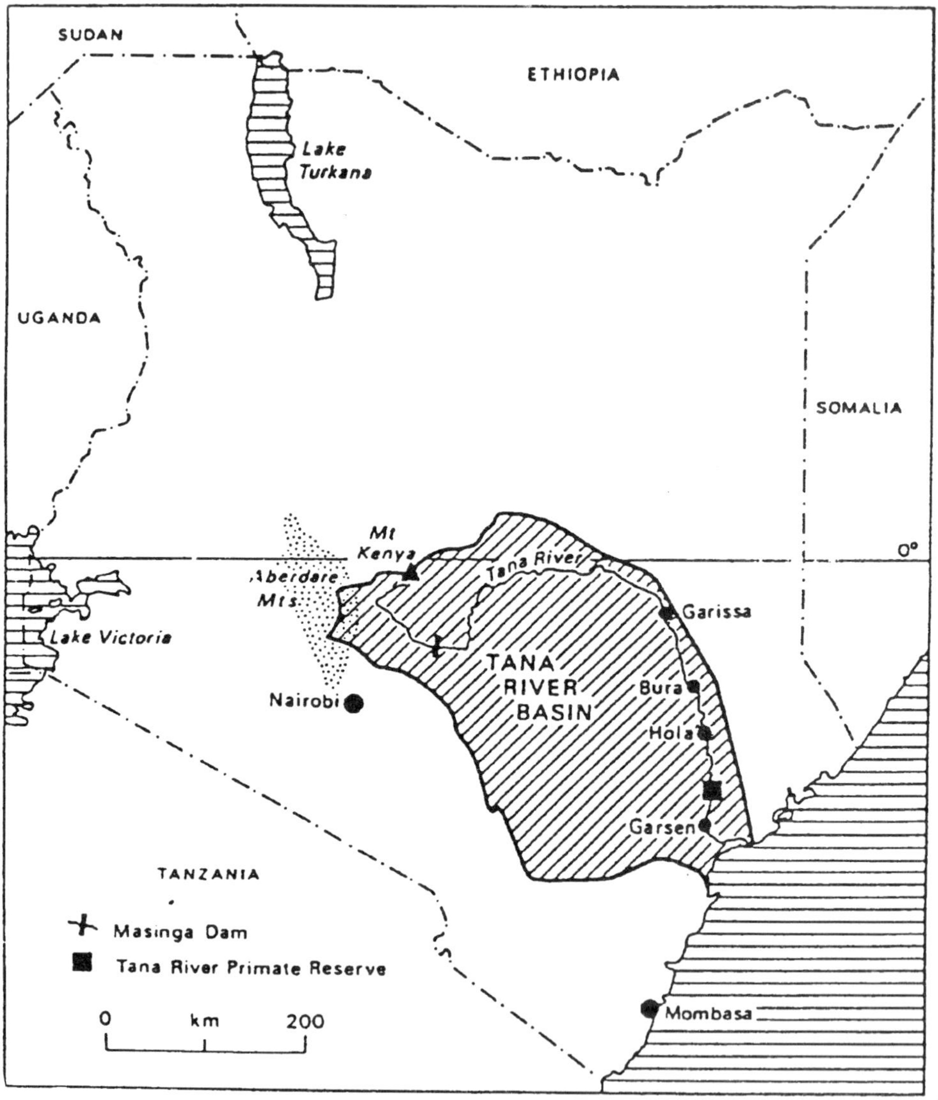

Map 1. *Map of Kenya showing the Tana River Basin* (cf. Hughes 1987)

They cultivate banana, rice, maize and vegetables for subsistence and also fish. Mangoes are collected and sold at the local markets and cotton has recently been introduced as a cash crop. Traditionally the floodplain forests, particularly the riverine forest, have belonged to them and only occasional dry season grazing by the Orma was allowed (Alakoski-Johansson, in preparation). The forest plays a central role in the life of the Malakote and provides the bulk of the materials used for house construction, household and agricultural implements, tools and weapons for hunting, foods, medicines etc. Forest products are also closely incorporated into social and religious traditions (Alakoski-Johansson, in preparation). The increased demand for forest-based products, due to the construction of the irrigation scheme, has added pressures on the forest resources.

The Orma, who number about 32,000 in total, are Eastern-Cushitic speaking pastoralists, who live in Tana River and Lamu Districts (Kelly, 1989). They raise in descending order of importance cattle, goats, sheep, camels and donkeys. Their territory in the semi-arid bushland away from the floodplain forest is roughly contiguous with the borders of the Tana River District. They distinguish three divisions which correspond approximately with the ecological zones and the subsequent modes of living. The most nomadic adaption required is in the area of the present Bura scheme. In the Galole area, south of Bura, the rainfall is higher and there are several locations, such as Waiyu, where permanent settlements have developed. The degree of sedentarization is higher among the Chaffa Orma in the Tana delta area and many of them do some farming (Ensminger, 1984). Recent land acquisitions for development efforts, such as the Bura and Tana Delta Irrigation Schemes (IBRD, 1990), proposed ranches and grazing block allocations (Kelly, 1990), and proposed land adjudication (BICEDA, 1988; Kelly, 1990), considerably increases the pressure on the Orma grazing areas and their life-style as a whole. As among the Malakote, plants, trees and shrubs, play a central role in animal husbandry as well as in the material, social and religious culture of the Orma (Alakoski-Johansson, 1991).

Development of the Tana Basin is concentrated along two different but not inseparable lines. One deals with hydro-power development of the upper reaches, and the other primarily with irrigation development of the lower basin. See Map 2. The effects of the former activities on the latter, including the ecology of the flood-plain forests, have been sufficiently demonstrated by Hughes (1987; 1990), Adams and Hughes (1986) and IBRD (1990).

UPSTREAM DEVELOPMENT OF THE RIVER TANA

The development of the Seven Forks Hydro-electric Complex along the Tana was dictated primarily by the need to reduce dependence on electricity imported from Uganda, other reasons were general growing demand for electrical power as population and urbanization increased. The current state of development of the hydro-power potential include dams at Kindaruma (completed, 1968), Kamburu (1975), Gitaru (1978), Masinga (1981), and Kiambere (1987). Another five dams, reaching as far downstream as the Kora Rapids near Garissa, are still in the planning stage. Ecological effects of dams are well documented. In order to mitigate the effects and prolong the lifetime of the investment, upstream watershed management as well as activities to reduce negative downstream effects are necessary.

The aim of the former is primarily to control soil erosion and subsequent siltation of the dams. The latter focus on reducing the ecological and socio-economic effects which are inevitably caused by the changes on the flow regime. Such mitigation measures may often be outlined in project feasibility studies and in project plans but are, however, too seldom implemented in practice.

The development of the hydro-power potential of the upper Tana is generally justified by national interests. It has taken place regardless of "minor" downstream effects, especially if these are of an environmental nature, but often also when it has been in conflict with groups which only have weak economic or political bargaining power. This development also conflicts with similar development efforts of considerable economic size and priority in the lower basin. The reason why this conflict was not recognized in two parallel efforts of such economic magnitude is difficult to appreciate!

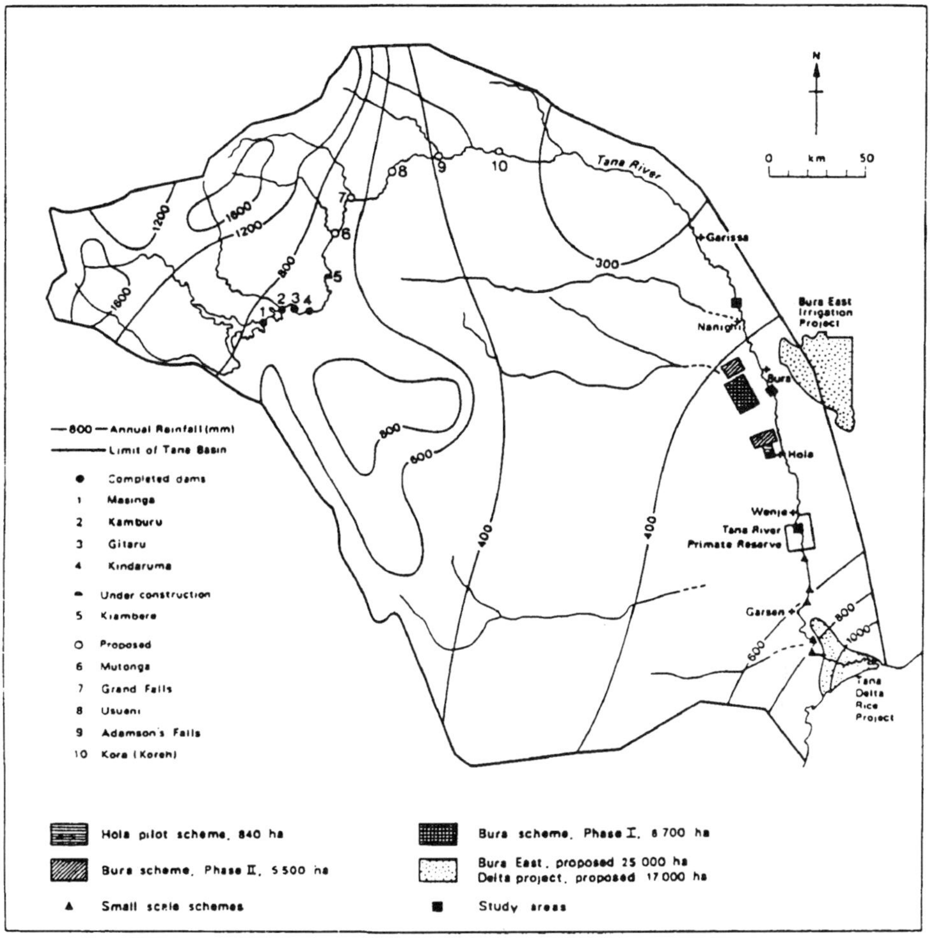

Map 2. *The Tana River Basin showing rainfall isohyets and some of the existing and proposed development schemes* (cf. Hughes 1987)

EFFECTS OF STREAMFLOW REGULATION

The effects of the dams on the river are mainly two-fold. Firstly, they regulate the discharge and reduce flooding. Secondly, the dams act as silt traps and may reduce the sediment load of the river. For instance, a simulation model of the discharge of the Tana showed a reduced frequency of high May flows due to the Masinga Dam of only 30 per cent over an average value, compared to 58 per cent without the dam. In March, the month of the lowest flow, the model predicted a considerably more constant and higher flow with the dam as compared to without (Adams and Hughes, 1986; Hughes 1990). It seems safe to assume that the dams will substantially modify the annual discharge regime and reduce annual peak flows. Ongwenny (1980, cf. Hughes, 1984a) estimated that the Kamburu Dam filled up at a rate of 4.2–5.1 million m^3 per annum and would last only 20–35 years, depending on the climatic trends. The trapping of silt and the reduced sediment deposition on the floodplain may also lead to degradation, accelerated erosion and reduced lateral river movement (Adams and Hughes, 1986; Hughes, 1990).

The downstream effects of the hydro-power development can roughly be divided into three main components; (1) effects on the floodplain ecology, (2) disruption of traditional farming systems, and (3) effects on irrigation development.

EFFECTS ON THE FLOOD-PLAIN ECOLOGY

The impact of flow regulation on the ecology of the riverine forest in the lower reaches is not fully understood, primarily because the successional characteristics of the forest or the factors affecting the forest structure, and the regeneration of different species in the forest are still poorly known (Hughes, 1985, 1988; Marsh, 1976; Medley, 1988). In her study of the Tana floodplain forest Hughes (1985) describes the roles of several factors with an impact on the extent, structure and regeneration of the forest. The flooding regime is an important factor together with the micro-topographical variations, which determine the distribution of water and sediment at flooding. These, in turn, are assumed to determine the species composition and structure of the forest and probably also the regeneration. Another decisive factor is the changing river course, caused by meanders being cut off. Hence, river control may have implications for the survival of the forest (Adams and Hughes, 1986; Hughes, 1990). How the reduction in flooding intensity and duration will affect the replenishment of the groundwater reserve, which is assumed to determine the width of the forest, is also not well known. The indications are that both the diverse structure and regeneration depend on occasional extreme flows, and that the present upstream development may threaten the existence of the floodplain forest (Hughes, 1985; 1990).

DEGRADING TRADITIONAL FARMING SYSTEMS

The Bura Irrigation Settlement Project during the years 1983–85 absorbed about 25 per cent of Kenya's total public investment in agriculture (IBRD, 1990), but little concern has been shown for the local population. Changes on the flooding regime have had direct consequences for the Malakote (and Pokomo further downstream) who depend on floods for their farming. For instance, the closing of the Kiambere Dam in 1987 led to an exceptionally reduced flow. The result was serious crop failures among the Malakote who were stricken by famine. These external forces, on the other hand, indirectly affect the forest by the pressure for alternative income sources. To compensate for decreasing yields from traditional farming practices, some Malakote turned to small-scale irrigation (probably inspired by the Bura Scheme), or to sales of fuelwood, charcoal and construction wood to the farmers of the neighbouring irrigation schemes. The need for more land, including intensified land use, may partly account for the increase in forest clearing, which e.g. has been observed at the Tana River Primate Reserve (Allaway 1974; Marsh, 1976; Medley, 1988). The clearing of forest is, however, in conflict with the policy of protecting indigenous forests so strongly advocated by the Head of State. This may trigger off conflicts over the use of the local resources. Are the rights to the resources to be determined by the traditions of the local community or by state authorities?

Charcoal manufacture and trade are examples of where the traditional system, which has not been prepared for the demands induced by external development, leads to resource waste due to lack of appropriate technical skills. The trade of fuelwood and charcoal to the nearby irrigation scheme has provided the Malakote with new opportunities. The Malakote are not accustomed to make charcoal, they simply set huge trees on fire, let them burn down slowly and collect the pieces of charcoal remaining. This is an extremely wasteful method even compared to simple earth kilns with low conversion efficiencies. In attempts to protect the forest, and thus follow national policy, local authorities prohibited burning of charcoal. Limiting rights and access to the forest is a potential source for conflicts, particularly when the problems are mainly caused externally by the state. The state has mainly offered restrictions, while blocking a local strategy to deal with the problems, instead of endeavouring to improve traditional farming practices. More frustration may well be caused by the fact that most attention in the district, including funds and extension work, has been directed to the new scheme rather than to the traditional population.

EFFECTS ON THE IRRIGATION SYSTEMS

The effect of the dams on irrigation development in the lower Tana Basin was not fully realized during the planning of the Seven-Forks complex. The effects were, however, clearly seen in 1987, when the river fell to a level that could barely sustain pumping at the Bura Irrigation Settlement Scheme. Water is the crucial question. The fact that without a gravity intake system the scheme was left to depend on pumps increased the risks considerably. Problems in water supply usually have direct repercussions on the farmers in the scheme through lower yield of cash or subsistence crops. The economic viability of the scheme depends on the output of the individual farmers. Hence, any activity which negatively affects their production is a loss for the whole enterprise and, eventually, to the state which has committed such considerable resources. Declining yields means declining incomes and purchasing power, which may force people to collect fuelwood from the natural forests. Moreover, insufficient irrigation water almost always excludes possible forestry components from the water distribution, and thus restricts the possibilities to mitigate environmental impacts by producing wood within the scheme. Due to such problems the management in Bura now considers impounding of water to be necessary (IBRD, 1990).

In summary, the upstream developments on the Tana have been dictated by national interests, chiefly of economic and political natures. These take precedence over most other aspects, including ecological considerations. The cause–effect relationships were not properly analyzed and negative impacts have not been mitigated. The ecological effects, both direct and indirect, threaten the existence of the floodplain forest and the socio-economic effects are felt downstream at district, village and household levels. Upstream dams have also partly contributed to the poor performance of the Bura Irrigation Scheme, in which such considerable political and economic interest was vested.

DEVELOPMENT OF THE LOWER BASIN: THE BURA IRRIGATION SCHEME

In 1977, the Government of Kenya, World Bank and other donors decided to finance the Bura Irrigation Settlement Project. See Map 2. This had been preceded by almost 50 years of plans and studies. The explicit aims of the project were to settle the landless, create employment, and contribute to foreign exchange earnings by producing cotton and maize for food (IBRD, 1990). It can be assumed that an implicit objective of the government was that the project be seen as an effort to bring the areas bordering Somalia, over which there had been violent disputes in the 1960s, closer to the central government and hence under a more strict control. In a larger context this should perhaps be seen as a step in a process of nation-building and unification, which may have been one of the reasons why the project took off regardless of the many attempts that were made to stop it on economic grounds during the early implementation (IBRD, 1990).

The original plans were to settle 5,150 farmers in 23 villages to cultivate cotton and maize on 6,700 ha of land developed for irrigation (IBRD, 1977). Approximately 4,500 ha of irrigated forestry plantations were proposed to cover the requirements of the estimated 60,000 people of the scheme. Since 1981 about 2,300 tenant farmers have been settled in ten villages cultivating 2,500 ha, which is only 37 per cent of the originally developed area. Land development, including clearing, was made for 6,700 ha. Also infrastructure and other developments were fully or partly developed to the original targets.

Bura was seen as only the first phase of the irrigation development of the west bank, which ultimately would join Bura with the pilot scheme in Hola. The Bura-Hola complex would comprise about 13,000 ha. Additional large-scale irrigation developments were foreseen on the east bank (about 25,000 ha), and further down in the delta about 17,000 ha were proposed for a rice scheme (Hughes, 1988). Currently the development of the rice scheme is well under way.

FORESTRY AND IRRIGATION DEVELOPMENT

Irrigation planning is mostly guided by the engineering approach, with little considerations to forestry or environmental aspects, even when it is obvious that neglecting these have negative consequences for the whole enterprise. Forestry components, providing fuelwood, shelterbelts, amenity trees etc. are often neglected in the planning or at implementation where agricultural, engineering and economic considerations always get the priority in decision making. Forestry related problems in connection with irrigation development are lack of fuelwood, building poles, trees for shelter, fodder and food. As usual it is the farmers on and off the scheme who suffer the unintended consequences.

The scheme has also created environmental problems. Irrigation schemes are frequently located in the arid and semi-arid areas, where the biomass production potential of the natural vegetation is low, and where the environment in general is rather fragile. Irrigation development inevitably leads to high population concentrations in areas with low carrying capacity. Because not enough attention has been paid to integrating forestry aspects, irrigation schemes often create development induced desertification of vast areas surrounding the schemes. This directly affects the schemes through wind damage, erosion and subsequent siltation of canals. The result is often serious land use conflicts with the traditional population of the area as well as degradation of the natural resources, which usually support the animal husbandry and subsistence agriculture of the indigenous peoples.

The need to include a forestry component, including forestry research, was realized at a very early stage of project planning in the 1960s (FAO, 1967). It was also included throughout the subsequent stages of planning, until it was formulated as a distinct, separate component of the project in 1982. The project document was finalized in 1986, after approximately 20 years of planning and at least nine separate planning, formulation and appraisal missions. Nevertheless, the final forestry component of the project was, although slightly more detailed in technical aspects, confined to a much more narrow scope than even in the first planning report (Johansson *et al.*, 1990).

In reality the forestry research started in 1984 as a small pilot component which was expanded in 1985. The actual forestry plantation component started in 1986, five years after the first settlers had arrived at Bura, and the first plantations were established in late 1986 and early 1987. By that time approximately 15–20,000 people were creating a new pressure on the forest resource and had removed an estimated 100,000 m^3 of wood from the area. This corresponds to a complete removal of vegetation from about 10,000 ha of bushland. Another 100,000 m^3 will probably be removed before the forestry plantations of the scheme start yielding sufficiently.

NEW DEMANDS ON FOREST RESOURCES

Estimates of population have ranged from 20,000 to almost 50,000 people. So it has been impossible to make estimates of consumption. In 1988 the District Officer at Bura estimated the population in the Division at 42,000 (Mwaura, 1988, pers. comm.). The number of tenant families is about 2,300 (Johansson *et al.*, 1990). With an average family size of 6.6 (Vainio-Mattila, 1987), the settler population would be about 15,000. A rough estimate of the satellite community would be about 4,000 people. Hence the local population would be about 23,000 people (Johansson *et al.*, 1990). Using traditional rule of the thumb estimates this corresponds to a fuelwood consumption of 42,000 m^3 per year.

A reasonable assumption would be that there was a balance in the supply and demand of forest resources prior to the establishment of the scheme. Hence the unstable situation has been primarily caused by the additional pressure created by the scheme. Fuelwood consumption estimates (measured) for farmers in the schemes have been made by Hughes (1982) in Hola and for Bura by Hughes (1984b) and Vainio-Mattila (1987). Similar estimates for the riverine villages have been made by Hughes (1982) and for the pastoral communities by Ensminger (1984). Very little is known about the quantities of wood used for other purposes. Some estimates of the amount of wood required for canoes have been made e.g. by Marsh (1976) and Hughes (1987). Nothing substantial, with the exception of the work by Ensminger (1984), has been presented on other uses such as building poles, which can be assumed to require the second largest amounts of wood.

A survey in Hola by Hughes of three villages in the Tana Irrigation Scheme (1982) estimated the annual per person fuelwood consumption to be 0.72 m^3. Corresponding estimates for Bura have been 1.13 m^3 (Hughes 1984b), and 1.23 m^3 by Vainio-Mattila (1987) in a survey of ten villages. The latter estimated a reduction in consumption of about 30 per cent using a Lorena-type stove instead of the traditional three-stone hearth. According to these studies the new settlement in Bura annually requires approximately between 18,000 and 24,000 m^3, depending on the number of fuel conserving stoves in use, their efficiency and the different consumption estimates.

A study of the regeneration potential and early development of *Acacia zanzibarica* bushland in Bura, conducted in 1988, concludes that the mean annual increment of the 6 years and 8 months old stand of the study area was 0.38 m^3/ha/year (Johansson and Kaarakka, 1990). The estimates are supported by a recent inventory (Pukkala, 1989), which shows that the annual growth in the bushland area is about 0.4 m^3/ha/year and in the floodplain area (outside the actual riverine forest) about 1.9 m^3/ha/year. Assuming that the annual demand created by the establishment of the scheme would be in the order of 18,000–24,000 m^3, and provided that this would be met from the natural forest, a sustainable utilization would require annually about 45,000–60,000 ha of bushland or 10,000–13,000 ha of floodplain forest. The obvious conclusion is that, due to distances and means of collecting, the wood will be collected from a smaller area resulting in local depletion of the woody biomass.

Studies in Bura have shown that the most promising species *Prosopis juliflora* is yielding about 10–15 m^3/ha/year under irrigation (Johansson *et al.*, 1990). Considering problems with water supply a conservative estimate of 10 m^3/ha/year would indicate a need of about 1,800–2,400 ha of irrigated forestry plantations in order to cover the new demand created by the Bura Scheme. Currently about 1,100 ha have been planted under irrigation. It has been estimated (Pukkala, 1989) that the plantations will reach an annual production of about 20,000 m^3 in 1994–1995, when the demand created by the scheme would be balanced. Very little is, however, known about how the removal of about 200,000 m^3 of wood, equal to about 20,000 ha of bushland, is affecting the ability of the forest to recover considering the simultaneous increase in grazing pressure and soil erosion.

CHANGES IN LOCAL UTILIZATION PATTERNS

Another observation from the studies in Bura has been that the riverine communities, which are surrounded by the most abundant source of wood in the area, are using more fuelwood (1.9 m^3/person/year) than those residing in the irrigation schemes (Hughes, 1982). The impact on the supply through possible degeneration of the floodplain forest, or the role of the increased demand to utilize the resource to cover decreases in food supply and the new market opportunities, are not known.

The study by Ensminger (1984) on the changes in the use of fuel and woodstock of the Orma reveals another interesting pattern. The degree of sedentarization seems to have a direct effect on the fuel consumption pattern. The study compared the fuel consumption patterns of three communities with different degree of sedentarization. The nomadic communities in the dry area with low woody biomass production potential used 0.46 m^3 per person per year. The communities settled about 50 years ago in the medium potential area consumed 0.74 m^3, which was about 60 per cent more than the nomadic Orma. Those who lived in communities settled about 100 years ago in the high production area (relatively speaking) used 0.56 m^3. The time expenditure for firewood collection in the different communities further clarify the picture. The time expenditure was 1.6, 21.2 and 48 hours per month per kitchen respectively. The pattern seems to go through phases from a situation of high-supply/low-needs (nomadic) to fair-supply/high-needs ("recently" settled) and further to a situation of low-supply/high-needs ("old" settlement). Similar trends were observed for other uses, such as building and fencing poles. The Orma at Bura are becoming increasingly sedentarized due to the services and job opportunities provided by the Scheme. It can be assumed that the same kind of increase in consumption of woody resources can be expected as in the cases studied by Ensminger (1984).

IRRIGATION DEVELOPMENT, LAND TENURE AND RESOURCE UTILIZATION

About 71,000 ha of land was gazetted for the Bura Irrigation Settlement Scheme (Idwasi, *pers. comm.*), most of this was Orma grazing land which was at least partly removed from the traditional livestock sector. Only a 3 km wide strip of forest along the Tana remained as Trust Land under the County Council. Furthermore, the development of about 17,000 ha at the Tana Delta Rice Scheme and other proposed developments (ranches, grazing blocks) south of Bura may restrict the traditional dry season grazing area of the Hiriman Orma from Bura.

A paper by Kelly (1990) indicates that there is a general trend to sedentarization by those Orma groups who traditionally have been nomadic. This, in combination with the limitations of grazing area, could lead to less mobile fallow herds and competition over wet season pastures with the already sedentary. She concludes, that the sedentarization of the remaining nomadic population would disastrously affect the general grazing and living conditions in the sedentary areas which are already limited. In Bura, the result would probably be more sedentary Orma settlements, less transhumant livestock, and more concentration of livestock on smaller areas. This would probably lead to more conflicts with the Somalis, who occasionally graze in the area during dry season, and possibly also with the Malakote who traditionally have only permitted occasional use of the riverine forest as a fall back grazing reserve (Alakoski-Johansson, 1991). From the ecological point of view the result would be local overgrazing of the bushland areas around the Bura Irrigation Scheme and increasing pressures to use the riverine areas for grazing. It can also be assumed that in such a scenario the regeneration of the indigenous forests would be endangered, and the ability of the forests to recover from more than ten years of depletion is questionable. Possibly the species composition of the surrounding areas would change, as it has in some areas around Hola, or the understorey would decrease considerably or partly disappear, as it has in some areas along the Turkwell in Turkana.

The need to conserve and protect the unique riverine forest has been included in the planning of the Bura Scheme. Ethno-botanical studies of the indigenous forests in Bura have revealed that the material and socio-cultural uses of trees formed one of the major foundations in both the Malakote and Orma societies (Alakoski-Johansson, 1991; Johansson and Alakoski-Johansson, 1988; Gachathi *et al.*, 1987). A total of 24,500 ha of natural forests were proposed for gazettement in three separate blocks, a provision also included as one of the loan conditions of the World Bank Second Forestry Project (IBRD, 1977). The conservation proposals have, however, been primarily ecological without giving consideration to the intimate relationship between the people of the area and the plants. The land tenure aspects have also been overlooked. As a result the traditional gazettement procedure, which requires the removal of people from the forest, was a difficult political problem in an area where already over 70,000 ha had been gazetted for other than the traditional population.

Proposals have ranged from traditional gazettement, to County Council Gazettement and allowing people to stay and manage the forest to supply wood for the Bura Scheme. The need for a management plan for the riverine forest has been expressed on many occasions and an attempt to produce a management plan for the riverine forest was made recently (Pukkala, 1989). But, the work remains more of an inventory than a management plan. It concludes that management should be aiming to conserve the forest. For whom, and by which aims and means remain unsolved questions. It seems likely that the difficulty in this particular case partly depends on an unwillingness to touch the sensitive issue of land and tree tenure, which is traditionally vested with the local people. Secondly, the problems seem to stem from the foresters inability both to appreciate the general complexity of such management and to see alternatives to the forest management procedures which have frequently been adapted from the temperate regions, where forests are treated more as units in physical planning than as basic elements of rural livelihoods and cultures, as is the case in Bura. Perhaps another approach to natural forest management, based on experience from Turkana, could have been developed. There, the rules for management and control have been based primarly on indigenous (Barrow, 1989). The problems of conservation and management still remain unsolved after more than ten years of discussions.

WATER MANAGEMENT OR A CRISIS OF MANAGEMENT?

One crucial mistake was made in the construction of the Bura Irrigation Scheme. Although the original plans for the water supply contained a provision for gravity intake, the decisions and the efforts to secure funding for the water supply system were always preceded by other urgent crises. Hence, the scheme remained with a temporary pumping station up to a point when the project had exhausted all credibility among major donors and the temporary system had become permanent. The heart of an irrigation scheme is the water supply system. What emerged during construction was a giant investment of about 105 million USD suffering from a heart failure. This was fatal for the agricultural scheme, because lack of water due to pump failures almost always occurred when irrigation demand was high. Consequently anticipated yields were never reached. For the Bura Irrigation Scheme the unreliable water supply was perhaps the single most important factor which turned the resource management strategy into crisis management and a struggle for survival.

For the forestry component the effects were primarily two-fold. Firstly, it turned interest away from anything else (including forestry) other than preventing the Scheme from sinking. Secondly, the unreliable water supply rigidly limited options for forestry development. One example is the selection of species. Some of the selection criteria were; (1) drought tolerance, (2) good survival at establishment, and (3) high yield at low levels of irrigation (Johansson *et al.*, 1990). The species in the pilot research which best fitted into this description was *Prosopis juliflora* but with some reservations, of which the most important was its tendency to become weedy (Johansson, 1985). Alternatives would have been species with a higher expected yield potential but which might be either wiped out when water supply was restricted, or continued experimentation.

The feedback from the choice of *Prosopis juliflora* has included several positive but also several negative aspects. On the positive side it has proved to be the fastest growing, most tolerant to a variety of situations, easy to establish and raise and with very high calorific value. Moreover, it has been easily accepted by pastoralists, who value the pods as fodder for animals, and it is widely planted by tenants as hedges and around their compounds. On the negative side currently there are indications that it is spreading along canals and into agricultural areas and may become a weed problem. Farmers who have planted it in combination with crops have been disappointed because of its competitiveness. So far farmers have been rather reserved about using it for fuel, partly because of the thorny provenances in which the first were planted, and partly because they are not familiar with it.

Another aspect of the water problem is that the scheme provided the forestry component with only limited quantities of water in 1986–1988. This diversified the strategy from primarily pure irrigated fuelwood production into exploring a variety of approaches (Johansson, 1986). The installation of new pumps in 1988 considerably increased the capacity, and water was released in sufficient quantities even for forestry plantations. Simultaneously the experience from rainfed plantations was rather disappointing. There were also difficulties in convincing the management, particularly the agricultural and irrigation engineers, of the need to incorporate trees to a higher degree along canals and other potential sites. Due to this the forestry component reverted primarily to planting irrigated areas only.

These were probably the main reasons for the rapid increase in forestry area under irrigation between 1988 and 1990. Recently problems in maintaining the pumps have re-emerged. This may force the forestry component of the scheme again into wider activities, as well as serious discussions with the scheme management on the role and status of forestry in the development of the irrigation potential.

CONCLUSIONS

The major factors affecting the ecology of the Bura area, either as direct or indirect ecological effects, are presented in Figure 1. They are broadly classified into effects caused by the development of the upper reaches of the Tana, those caused by the development of the delta and the lower reaches of the Tana, and by effects of local development in Bura. The local effects are further separated into internal effects within the scheme and those which relate to the interaction of the irrigation scheme with the division or district. In the local effects some positive as well as some negative aspects are mentioned.

The political and economic interests were too strong to stop the construction of the Bura Irrigation Settlement Project, regardless of serious doubts about the economic viability of the scheme. Although some of the ecological implications were seen at planning, many of the impacts in the complex network of effects caused by the development of the Tana Basin were either not sufficiently studied or entirely neglected. And in those cases where mitigation was proposed, such as in environmental and forestry issues, at implementation other aspects always received higher priority. The result will be considerable environmental degradation of the Bura area for at least 15 years.

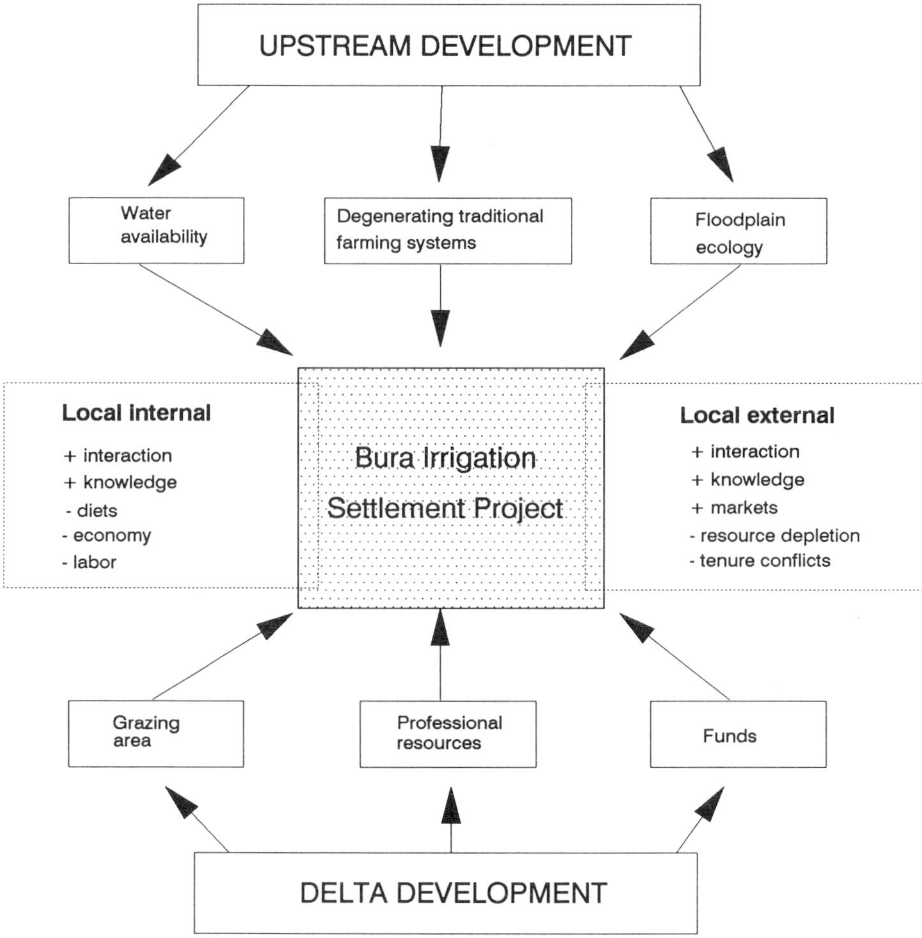

Figure 1. *Some factors affecting the ecology in the Bura area and the management of the Bura Irrigation Settlement Project*

The two focal areas of development (upper Tana and the lower basin) were treated as separate issues and their inter-dependencies were not appreciated. An example of this is the impact of the dams on the water availability, floodplain ecology and traditional riverine farming practices, which all have serious implications for the Bura Irrigation Scheme.

Another example is the development of the irrigated rice scheme in the Tana delta. One can well ask how a new economic investment in irrigation is justified, when the previous one in Bura has a negative economic rate of return and requires about 1,000 USD in annual subsidies per settler (IBRD, 1990), which will at least have indirect impacts on Bura. It is likely that it will burden the already strained government budgets, as well as the limited professional irrigation expertise in Kenya. Secondly, the decision to establish the delta scheme seems to have been taken without considering the impact this has on Orma grazing conditions. The subsequent increase in grazing pressure and the possible repercussions on the re-generation and general ecology of the Bura area were not seen as a consequence of the delta development.

Another aspect of the development at Bura is that the irrigation project was treated in most cases as a separate entity, without linkages to the overall situation in the division or district. Most planning reports and studies only deal with the scheme. An example of this is the lack of population estimates, which is the basis for all energy planning, particularly in a case where most of the fuelwood comes from outside the scheme. Another example is the recent World Bank completion report (IBRD, 1990), which in its analysis of the Bura Irrigation Scheme barely touches the issue of the impact of the scheme on the surrounding population or on the environment. This seems rather surprising at a time when environmental questions are treated as key issues in most development discussions. Alternatively one can see it as another indication of the general discrepancy between the catchword loaded policy and strategy jargon of development banks and agencies and the crude reality of everyday implementation, where these words do not yet seem to have taken root.

If environmental considerations carry any weight in irrigation development then forestry has to become an integral part of both planning and implementation, particularly in arid and semi-arid areas. This has to be reflected in institutional and organizational set ups, land and water allocation and budgetary allocations for the operations and maintenance required for forestry. This should also be reflected in the decision making priorities. Currently this is rarely the case. In Bura the forestry component has largely been seen as an external activity attached to the irrigation project, not necessarily as a part of the day to day management. Environment Impact Assessment (EIA) procedures should be adopted at the project preparation stage and operational alternatives should be presented where major cause–effect networks have been outlined. More strict project monitoring procedures, including links and arrangements with the operational management, should be developed. In cases such as Bura the environmental problems are primarily caused by the strong overall increase in population as well as in population density, although external influences have also contributed to the rate at which environmental deterioration has taken place. The obvious solution is that the problem should be solved where it is caused, i.e. primarily within the schemes.

Research at Bura shows that the backbone should be irrigated forestry plantations, since a sustainable utilization of the indigenous forests would require such vast areas that management alternatives would be very difficult or impossible in practice. The design, allocation and management of such plantations depend on the flexibility of the original land development and irrigation works. This is the prime reason why forestry aspects should be included from the first stages of planning. In Bura the design, although it contained provision for about 4,500 ha in huge blocks, was such that it allows little flexibility for other than the centrally managed block option. There should be flexibility to allocate land for trees to the villages for farmer or village managed forestry. The use of drain and seepage water for forestry can considerably improve water use efficiency. For large settlements the use of sewage effluents and sewage treatment facilities is another way to more efficient use of available water resources, although more research is required on the possible accumulation of potentially unhealthy substances particularly in fuelwood. By developing the drain designs, and also by water harvesting techniques, surface water can be harnessed for productive purposes. Also the design of the canals which provide farmers with water for their compounds should be included in the irrigation works. Neglect of this may lead to the use of purified drinking water to irrigate trees in the compounds, as has been the case in Bura.

The case of Bura and the whole development of the Tana Basin shows that the ecological effects of development efforts are rarely properly analyzed in advance and frequently not even in ex post evaluations. The complex network of effects, both social and ecological, caused by an intervention in an ecological entity, such as a catchment area, is rarely sufficiently mapped. It seems that the hierarchy of interests, primarily of economic and political natures, is blindfolding, i.e. neither the potential kickbacks nor the original causes are recognized through the complex network of effects or hierarchical levels. Moreover, it seems that the development of large-scale agricultural irrigation projects are frequently guided by crisis management rather than by management of resources.

Although it may not be possible to see the whole picture in advance, many of the ecological cause–effect relationships can be depicted and prioritized in advance. Based on the Bura experience it seems that such problem analyses are too seldom made and even more seldom brought into operational management alternatives, but they are necessary in order to provide a range of scenarios on which environmentally sound alternatives can be based.

REFERENCES

Adams W.M. and F.M.R. Hughes, 1986: "The environmental effects of dam construction in tropical Africa: Impacts and planning procedures". *Geoforum* ,Vol. 17, 3:403–410.

Alakoski-Johansson G., 1991: *Traditional uses of trees and shrubs among the Malakote and Orma in Bura, Tana River District,* Kenya. (Forthcoming).

Allaway J.D., 1979: *Elephants and their interactions with people in the Tana River region of Kenya.* PhD-thesis, Cornell University.

Armitage F.B., 1985: *Irrigated forestry in the arid and semi-arid lands: a synthesis.* IDRC-234e, the International Development Research Institute, Ottawa.

Barrow E.G.C., 1989: "Trees, people and the drylands: The role of local knowledge". In A.M. Kilewe, K.M. Kealey and K.K. Kebaara (eds), *Agroforestry Development in Kenya.* Proceedings of the Second National Seminar on Agroforestry, Nairobi, Kenya, 7–16 November, 1988. ICRAF. Nairobi.

BICEDA, 1988: *Bura Fuelwood Plantation Project Phase* 2. An appraisal of BFPPs Plan of Operations and recommendations for a strategy towards a sustainable fuelwood supply in the Bura Division. The Beijer Centre for Energy and Development in Africa, Nairobi.

Ensminger J., 1984: "Monetization of the Galole Orma Economy: changes in the use of fuel and woodstock". In C. Barnes, J. Ensminger and P. O'Keefe (eds), *Wood, energy and households, perspectives on rural Kenya. Energy, environment and development in Africa,* the Beijer Institute and the Scandinavian Institute of African Studies. Uppsala.

FAO, 1967: *Kenya: Survey of the irrigation potential of the lower Tana River basin.* Food and Agriculture Organization of the United Nations, Rome.

FAO, 1985: *FAO production yearbook.* FAO. Rome.

FAO, 1986a: *Databook on endangered tree and shrub species and provenances.* FAO Forestry Paper 77, FAO. Rome.

FAO, 1986b: *Irrigation in Africa south of the Sahara.* FAO Investment Centre Technical Paper 5, FAO. Rome.

Gachathi F.N., G. Alakoski-Johansson and S. Johansson, 1987: *A check-list of indigenous trees and shrubs of Bura, Tana River District, Kenya with Malakote, Orma and Somali names.* Bura Forestry Research Project Technical Report No. 6, Kenya Forestry Research Institute and University of Helsinki.

Hughes F.M.R., 1982: "Provision of fuelwood for irrigation schemes: the case of the Bura Irrigation Settlement Project, Kenya". In H.G. Mensching (ed). *Problems of the management of irrigated lands in areas of traditional and modern cultivation.* IGU Working Group 1, Egypt Meeting Report 1982.

Hughes F.M.R., 1984a: "A comment on the impact of development schemes on the floodplain forest of the Tana River of Kenya". *The Geographical Journal,* Vol. 150, No. 2, July 1984:230–244.

Hughes F.M.R., 1984b: "Fuelwood: a forgotten dimension of irrigation planning". In W.M. Adams and A.T. Grove (eds) *Irrigation in tropical Africa.* Cambridge African Monograph 3.

Hughes F.M.R., 1985: *The Tana River floodplain forest, Kenya: ecology and the impact of development.* PhD-thesis, University of Cambridge.

Hughes F.M.R., 1987: "Conflicting uses of forest resources in the lower Tana River basin of Kenya". In D. Anderson and R. Grove (eds) *Conservation in Africa: policies and practice.*.Cambridge University Press.

Hughes F.R.M., 1988: "The ecology of African floodplain forests in semi-arid zones: a review". *Journal of Biogeography,* 1988 15:127–140.

Hughes F.R.M., 1990: "The influence of flooding regimes on forest distribution and composition in the Tana River floodplain", Kenya. *Journal of Applied Ecology,* 1990, 27:475–491.

IBRD, 1977: *Bura Irrigation Settlement Project appraisal report.* Report No. 1446-KE, World Bank. Washington D.C.

IBRD, 1990: *Project completion report. Kenya: Bura Irrigation Settlement Project.* Report No. 8493, IBRD. Washington D.C.

Idwasi, 1988: General Manager, BISP, pers. comm.

IUCN, 1978: *Red data book,* Vol. 1, Mammalia. IUCN, Gland.

Johansson S. 1985: *Forestry development in the Bura Irrigation Settlement Project.* Final report for 1984. University of Helsinki, Department of Silviculture.

Johansson S., 1988: "Alternative afforestation strategies for the Bura Fuelwood Plantation Project". *Bura-FORP Working Paper 18.* University of Helsinki and Kenya Forestry Research Institute.

Johansson S. and G. Alakoski-Johansson, 1988: "Etnobotaaninen tutkimus tropiikin maaseudun kehittämisessä: Kokemuksia Buran metsäntutkimushankkeesta Keniassa" (with an English summary). *Silva Carelica* 12:263–269.

Johansson S.,V. Kaarakka, O. Luukkanen and J. Mulatya, 1990: "Forestry in irrigation schemes I. Research activities at Bura, Kenya 1984–87". *University of Helsinki Tropical Forest Report* 4.

Johansson S. and V. Kaarakka, 1990: "Regeneration potential and early development of Acacia zanzibarica bushland in eastern Kenya". Paper presented at the research seminar on Natural Dynamics and Regeneration Methods in Dry Tropical Forests, University of Uppsala, Uppsala, 9–12 October, 1990.

Kelly H., 1990: "Commercialization, sedentarization, economic diversification and changing property relations among Orma pastoralists of Kenya: some possible target issues for future pastoral research". In P.T.W. Baxter and R. Hogg (eds) *Property, poverty and people: changing rights in property and problems of pastoral development,* Dept. of Social Anthropology and International Development Centre, University of Manchester.

Mac Donald Sir M. and Partners, 1977: *Project Planning Report, Bura Irrigation Settlement Project.* A joint consultancy report by MacDonald and Partners, Huntington Technical Services, and East African Engineering Consultants for the National Irrigation Board, Government of Kenya.

Marsh C.W., 1976: *A management plan for the Tana River Game Reserve,* New York Zoological Society and University of Bristol.

Medley K.E., 1988: *Forest structure, composition and dynamics relative to primate populations on the Tana River, Kenya.* Six-Month Progress Report, Wildlife Conservation International, New York Zoological Society.

Mwaura, 1988: District Officer, Bura, pers. communique.

Ongwenny G.S.O., 1980: "The Tana drainage area". In *Water development and the environment in Kenya.* UNEP/UNDP/Government of Kenya Project.

Pukkala T., 1989: *Inventory and management plan for the indigenous forests and fuelwood plantations of Bura Irrigation Settlement Scheme, Kenya.* Technical Report No. 1. Bura Fuelwood Project, FINNIDA and the Ministry of Environment and Natural Resources, Kenya.

Tana and Athi River Development Authority, 1980: *The Tana Irrigation and Reservoir Scheme.* Government of Kenya, Nairobi.

Vainio-Mattila A., 1987: *Bura Fuelwood Project. Domestic fuel economy.* Report 13 B, University of Helsinki, Institute of Development Studies.

Are Trees Effective Against Desertification? Experiences from Niger and Mali

Michael Angstreich

This paper emanates from a presentation accompanied by slides on agro-environmental problems in the Sahel. Therefore it does not take the usual paper format. However, the main purpose is to introduce experiences with different forestation technologies such as shelter-belts and sand dune stabilization and their capacities as possible solutions to the ecological crisis in the West African Sahel.Two case studies are presented: Majjia Valley in Niger and Koro, Macina and Timbuktu projects in Mali.

CASE 1. MAJJIA VALLEY IN NIGER

The Majjia Valley lies in south-central Niger of which seventy-five percent lies within the Sahara desert. The soils of the valley floor are moderately fertile over a shallow water table and cover an area of 450 square kilometers. As late as the 1950s there were trees, bushes and wildlife in the area; today most looks like a moonscape. Average rainfall is about 450 mm per year with 50 per cent variations from year to year. The rainy season lasts from May/June to September, i.e. about five months. From November to April the Harmattan wind blows continuously from the Sahara, causing a great deal of wind erosion. Winds from the opposite direction during the cultivation period also cause serious damage to crops. The movement of sand dunes located just north of the valley threaten agricultural land.

It is commonly appreciated that trees prevent topsoil from being blown or washed away and protect agricultural land against the damaging effects of too much wind, water and sun. They can increase the soil's ability to retain moisture and enrich it with organic material. Trees produce fuelwood, construction materials, browse, fodder, fruits and medicines, not to mention cooling shade in hot climates.

Niger suffered serious droughts in 1973/74 and 1981–1984. The earlier crisis contributed to a change in the country's development process when a military coup brought in a government which gave high priority to ecological rehabilitation and food production.

For generations the agro-environment of the Majjia had not been overly threatened. With long fallow periods trees survived as an important component of the area's shifting cultivation, regeneration of natural vegetation took place and the soil became rejuvenated. Because of increased population growth, close to three per cent in recent times, this is no longer the case. Demands for pasture, fuelwood and especially agricultural land have increased in step with population growth. Trees became victims of this growth. As trees disappeared the soil, which people were dependent upon, became vulnerable to wind and water erosion. Grain harvests diminished considerably. Sand dunes threatened agricultural areas north of the valley. About 33,000 Azna-Hausa people live in 40 villages throughout the valley.

The majority of people in the Majjia live directly from agriculture. They cultivate millet, sorghum, beans, cotton, melons and ground nuts. The men identify mostly with the cattle, while the women have flocks of sheep and goats and raise livestock for food and for long term investment. They need fuelwood for food preparation but, because of the lack of wood, increasingly use millet stalks for cooking. Earlier, the stalks would have been used for construction purposes or animal fodder. Similarly, people have begun using dried animal dung for fuel instead of as manure.

Increasing population pressure has forced a drastic reduction in, or even elimination of, the fallow period. Statistics from Niger show a 40 per cent decrease in millet yields per land unit since the 1950s. Rural inhabitants tell of villages that were surrounded by trees a generation ago which have disappeared. Women spend many hours a day collecting fuelwood and water.

Much of the Majjia Valley is seriously degraded due to the lack of vegetative protection against wind and water erosion. The valley is especially prone to strong winds which funnel through it for most of the year. In 1974, a local agent of Niger's Forest and Water Service, CARE and the local population started efforts to stop the degradation process in the valley. It is important to note that the people themselves actually initiated efforts and asked the officials and CARE for technical support.

Sand dunes

Sand dune stabilization activities began north of the valley. Sand had already covered one village and threatened agricultural land. After the grain heads have been harvested millet-stalk fences are constructed across the dunes. These reduce wind speed and sand movement enough to allow wild seeds and planted trees to take root and grow. Between 90 and 100 hectares have been stabilized with millet fencing, tree planting and natural regrowth. The cost per hectare was about USD 7,000.

The valley floor

Results from the valley itself show that, for about USD 280 per hectare, cultivated land can be protected from wind erosion and sand concentration with shelter belts. Before the project began serious crop damage due to strong winds was common. Local farmers tell of having to replant their fields several times and still losing everything. CARE provides project financing, seedling bags, fencing and technical advice. Villagers provide land, labour and, not least, their knowledge and skills. Double rows of either all neem or neem/acacia trees are planted by the villagers.

Today, 4,600 hectares, about ten per cent of the valley floor, are protected by approximately 460 kilometers of windbreaks. Each double-row belt of trees is planted perpendicular to the prevailing wind direction. Visits for farmers from other areas are organized regularly and there has been some limited, spontaneous replication elsewhere.

Research results

A number of research efforts indicate that the tree lines reduce windspeed by from 40 per cent to 80 per cent, that there is improved retention of soil humidity and that grain yields within the tree-protected area increased by 15–20 per cent. Investigations also showed that production of millet bio-mass within the tree-protected area was 68 per cent higher than bio-mass production outside. This is especially important in the local context because the entire millet plant is used in several ways: as animal fodder, for fencing, as construction material, as fuel, etc. Tests also showed that while chemical fertilizers increased grain yields they did not increase bio-mass significantly.

More recent studies show that a full pollard cut, i.e. full crown removal, on one line of trees can be carried out without significantly reducing the other line's protective effect. These studies also indicate that ten kms of eight year old trees, the earliest age at which trees should be pruned, can provide enough fuelwood to satisfy ten per cent of the valley's population for one year. After the 8th year, trees can be pruned every 4th year.

Nurseries

Three types of nurseries provide seedlings to the Majjia Valley: private mini-nurseries producing up to a few hundred trees, village nurseries producing up to a few thousand trees and central nurseries producing several thousand trees. The project is encouraging more production from the private nurseries in an effort to reduce costs. Tree-seedlings are planted out by men, women, school children and youth clubs in the beginning of the rainy season. The most important species used are *Azadirachta indica* (neem), acacia and prosopis. Neem was brought to West Africa from India in the 1930s. It is fast-growing and shady and is used in the higher row of trees within each shelter belt. Neem produces fuel-wood, forage, tooth picks and seeds from which an insecticide (nimocine) can be pressed.

Acacias are leguminous trees which can provide nitrogen to the soil. *Acacia nilotica* is planted for the shorter, windward row of each belt of trees, its leaves and pods provide animal fodder and compost, while the thorns act as a deterrent to livestock browsing. *Acacia albida* drops its leaves in the rainy season, going into a sort of dormancy period, thereby reducing competition with millet for sunlight and nutrients. *Prosopis julifora* is a thickly growing thorn-tree which is planted along stream banks to provide protection against water erosion.

Protection

For trees to protect the soil they need to be protected against livestock. Several options were tried. Fencing of individual trees and of entire fields was too costly. The most effective approach was to hire and train local watchmen, who patrol and explain what they are doing to those they encounter planting new fields. The protection of young trees lasts from 2–4 years, until the leaves grow out of reach of livestock. Protection represents a major portion of project expenses.

Participation

A key to positive results is local participation in all phases of the project. Only active participation from the people themselves can assure continuing progress in the protection of their own resources. A survey in 1985 showed that the great majority of the people who live in the valley are convinced that the windbreaks are to their advantage. But we have had certain problems.

Ownership of the trees needs to be clarified. The majority of the people believe that the government owns the trees, even though the government recognizes usufruct rights of the people. This confusion may hinder a faster spread of the project to other areas. The problem is being tackled by regular discussions between local project participants and officials, with the latter guaranteeing ownership to those who plant and protect trees.

Women's participation has been weak in the past due to both traditional practice and to unawareness on the part of the project leadership. While trees are being planted and in the course of the first years, livestock, mainly the goats and sheep which women control, need to be kept away from planting fields where they would normally graze. Now, women are actively consulted before planting begins and alternative grazing and fodder supplies are planned ahead of time.

In summary, micro-climatic conditions in the windbreak area have improved and windspeed and wind erosion have been reduced. It has been shown that windbreaks can contribute to increased grain and bio-mass production. A rotation-system for the pruning of trees for fuelwood, poles and fodder, without markedly reducing the effect of the windbreaks, looks very promising. Local participation needs to be strengthened.

Is the windbreak-method appropriate for the Majjia Valley and can the project be replicated in other areas? The windbreaks have had a positive effect for the Majjia Valley and its inhabitants. But we now believe that this is not the only method which should be promoted and we are looking at the possibility of dispersing acacias on cropland, the protection of natural regrowth, grazing management (not least for manuring fields), terracing and other soil conservation techniques for the valley's slopes, more work with soil conservation along stream banks, fruit trees, woodlots and other activities which will form a more holistic approach to the agro-environmental conditions of the area. We know from our experience that we cannot replicate the project just anywhere in the Sahel. However, the Majjia is one of several ancient river valleys in Niger and other Sahelian countries which have relatively high water tables and moderately fertile soils. These valleys are centres for food production and settlement. To protect these areas and recreate sustainable farming systems will be an important contribution to the future welfare of the Sahel.

CASE 2. KORO, MACINA AND TIMBUKTU PROJECTS IN MALI

Koro, Macina and Timbuktu projects encompass several components such as gardening, agro-forestry, water management/harvesting, improved stoves, training, etc. Each project, however, has its own special emphasis. In Koro it is agro-forestry, in Macina it is gardening and in Timbuktu it is water management. The projects evolve according to experience and the emphasis is subject to modification. In this section I limit my remarks to water harvesting in the three project areas.

The annual rainfall in Koro and Macina is about 600 mm and between 170 mm and 250 mm in Timbuktu. The rainfall figures in Timbuktu (Dire and Goundam cercles) may be misleading since the farmers depend on Niger River flooding for cultivation. The participants in Koro are Dogon millet farmers, in Macina are Bambara grain farmers and in Timbuktu are Songhai riverine farmers.

Something old that is still going on

In Timbuktu our work is based on adaptions to traditional water management systems. We work with communities which cultivate depressions which become flooded by the surging Niger river. In these flooded areas, called *mares*, the people grow rice and vegetables. The traditional dikes are built low, compaction is minimal, and the dikes are therefore prone to serious erosion by the force of the flooding river. The traditional clay, wood and sandbag sluice gates provide only minimal water control. It is not uncommon to lose entire rice crops in the immature stage due to overflooding, or to drying out at later stages when water cannot be retained properly in the "mares". By building higher, more compact dikes, with cement-framed sluice gates, intake and discharge of water can be controlled according to the needs of the crop at its various stages of growth. The agricultural season can also be extended by holding back receding waters. This "recession agriculture" has been especially beneficial for, but not limited to, producers of vegetables.

Dikes may extend for hundreds of meters, and have a height of 2–3 metres and a wall-slope to a height ratio of 2:1. Top width is about 2 meters. We get design and evaluation help from Abdullai Diawara at Mali's National School of Engineering. The dikes serve cultivable areas of 50 to 400 hectares. The work involved in building a dike is extremely hard, especially in the February–May period, the hottest part of the dry season, when most work is done. Food-for-work is normally provided for dike construction. Workers from the area use picks to break up the hard clay soil, then pulverize it and carry it to the dike line. There it is laid in 10–

25 cm layers which the people compact with heavy pounders until the 2–3 metre height is attained.

Since 1986 improved dikes and sluice gates have been built by thousands of people covering thousands of hectares. These structures have had a positive impact. Average yields for deep-water rice on non-managed flood plains is about 1 ton/hectare. The average yield using improved dikes and sluice gates is 2.7 tons/hectare, or nearly a three-fold increase. According to a project test, yields could reach over 4 tons per hectare under ideal water management conditions without any fertilizer use.

Maintenance is still a concern even with the improved dikes. The wall facing the river is prone to erosion from waves and currents. Several perennial grass species are being planted on the walls and tops of dikes as natural protection from wave action. One of the species is *bourgou* (*Echinochloa stagnina*), a valued forage species. Others are *Eynodon dactylon* and *Vossia cuspidata*. About 4,500 m^2 of grass-plantings are being tested at present. Village groups are responsible for the annual maintenance of dikes.

Away from the flooded areas, another water harvesting technique is being used, the *zay*, or shallow planting basin, which collects scarce rainwater, and in this area it is truly scarce; only 170–250 mm per year. The *zay* has nevertheless proved to be effective for planting trees. We experienced an 80 per cent survival rate last year in a test in which no watering was done. This is an excellent survival rate in even the best of conditions, which the Timbuktu area clearly does not provide.

Before closing with this example, I should mention the cross-fertilization that takes place between this project and the one run by Norwegian Church Aid in the Gossi-Gourma area, several hours to the east by car. NCA is engaged in similar activities and we have reaped the benefit of staff exchanges, visits and co-run seminars.

Something new that went wrong

The Macina area is not far from Segou and the vast ricefields of the Office du Niger. As one of several activities, we thought we would encourage vegetable gardening in the height of the dry season, i.e.

February–April, also known as the hot season. Wells would be built for the primary purpose of supplying water to gardens established around each well and cultivated by villagers on a rotational basis.

Aside from technical problems, several social issues arose because of the wells:

1. The wells were used by gardeners and also by the rest of the village to draw water for various purposes. When repairs were needed, as invariably happens with wells, persons were to be assigned to assist project technicians in carrying out the repairs. Disputes arose over which villagers should do the work; should it be the gardeners who had plots around the wells or other villagers who used the wells only to gather water for household use? Each claimed the other was responsible and a number of wells became inoperable. Even before this it became apparent that certain wells could not cover the needs of both gardeners and the rest of the village.
2. The wells were constructed by the community on public lots, with gardening a pre-condition for building a well. From the start the project asked that eight gardeners be assigned to each well, and that they cultivate a certain number of beds covering a certain area with specific vegetables as determined by project staff. A rotation system was planned whereby every villager who wanted to would get a chance to garden. The selection of gardeners was left to the village chief and counsellors.

However, many of the first gardeners were unwilling to give us their gardens when the time came because they had put in a lot of work establishing the garden beds, building dead and live fencing and even planting some trees. It should be noted that during the height of the dry season the gardens are the only greenery around and act like a magnet for the local insect and rodent population, not to mention village goats. The gardeners formed watch groups to tackle this problem.

To resolve the more important social issues, it was decided by staff and villagers that:

a) The primary purpose of new wells is to solve the problem of water shortage for general household use. Once the water supply to the village as a whole is no longer an issue and a reason for dispute, gardening activities may be encouraged.
b) If a village decides to have gardening activities in connection with a well, the size and number of gardens, as well as what vegetables to produce, will be left up to the village to decide. CAREs extension agents are available for technical advice.

Something new that went right

The village of Pondourou lies in the southern part of Koro Cercle near the border with Burkina Faso. One of our Malian extensionists, David Dara, has been working with the farmers of Pondourou since 1988 to protect their agricultural land and to improve food production. The village is surrounded by eroded land. It has experienced serious reductions both in grain yields and in the amount of land which can be farmed. David Dara organized a simple demonstration for the local people. He climbed a mound of bare, eroded soil with a watering can and simulated rainfall. The water ran down the slope carrying soil particles without infiltrating the soil surface. It was obvious to the on-lookers that water run-off was of little value to their agricultural crops. David then told how "diguettes"—low stone barriers laid on the contour —and planting basins called "zai" (something new David had learned about on a visit to Yatenga in Burkina Faso), could retain water and protect and rejuvenate soils. The villagers decided to give them a try.

They chose a nearby piece of land which was considered completely useless. David Dara demonstrated the use of a water-level made with a piece of hose with which the people could determine the contours of this field and those of their own. In the course of three weeks the people, during their free time, laid three diguette lines and covered the field with small zai in which millet was planted. The rains came and the structures generally held their ground. Water and soil erosion was limited. Water infiltration was increased. The millet that was planted on this so-called useless and abandoned piece of ground germinated well before the millet in neighbouring cultivated fields. The final harvest was also much better.

These results made a strong impression on the people of Pondourou, several of whom decided to employ the same techniques on their own fields. One farmer also remarked that this was not only cheaper than paying for Forest Services permission to clear new land for farming, but he could farm for longer periods closer to the village.

We do not have statistics yet, but those from the Yatenga Project (Agro-forestry in Africa, Panos 1990, London) indicate up to 90 per cent yield increases. This is one example of how simple technology, in this case introduced from a neighbouring country, has made a positive contribution to improved production and environmental rehabilitation.

Nevertheless we must continue to promote effective, appropriate water harvesting techniques; we can still wind up where we started or even worse, unless some answer to the locust problem in Mali is found. We visited the Koro area last year, saw some impressive soil, water and tree conservation work which had taken years to accomplish in the project, but also saw that, in a few short weeks, locust/grasshoppers had destroyed 50 per cent of the millet crop in the area.

Community-Based Sustainable Development in Central Butana, Sudan

M. E. Abu Sin

This paper reviews the patterns and trends of development in the semi-arid lands of the Sudan, with special reference to Central Butana. It also seeks to examine how far it is possible to embark on community-based sustainable development as an alternative to top-down development. To realize this objective the paper describes the 1988–1990 community-based development project in Central Butana, which sought to effectuate a bottom-up approach based on social animation, as a means of involving peoples in community development planning and implementation. Within this context I examine the successful, self-generated, adaptive and adjustment response pursued by the community, which, I believe, is the only line of action for sustainable development.

AN OVERVIEW OF DEVELOPMENT OPTIONS

In Butana, as in so many areas of the Sudan, development has neither been designed nor geared to improve the life of the local inhabitants. The primary form of development has been to provide a water supply in the forms of machine constructed *hafirs*, boreholes etc.; these have been seen as the base for other services such as education and health.

Large-scale gravity irrigation projects and rainfed crop production farming have been encouraged on the edge of the study area, which have accommodated some residents as tenants. These enterprises have been initiated, funded and managed either by Government or private investors, and they contrast in nature, structure and management with the traditional production systems. Self generated development in response to contact with such enterprises and urban centres is manifested in the forms of changing herding strategies (Abu Sin, 1982), acquiring tenancies, and picking up outside jobs and businesses. The result has been imbalanced use of resources, sharp land use conflicts and a highly competitive state in which the majority of the Central Butana people have been the least equipped and powerful. The natural victims have been the majority of the population whose adaptation capabilities are limited.

In order to have a better background to the community-based option advocated, it is necessary to look first into the top-down option. The attitude of the community towards services has been positive for the following reasons:

a) the pressing need for such services specially water supply, education, etc. and the benefits families have acquired in the process of breaking their isolation.
b) the primary services have been compatible with some of the peoples aspirations, and the community has managed through self-help programmes to participate in the rehabilitation of services.
c) they have provided lessons on how to tap central resources which may help in mobilizing people in the direction of community-based productive enterprises.

The large-scale projects of modern agriculture, in spite of their national rather than local objectives and their rigid planning model of development, have had the following positive aspects in paving the way for community-based development. Physical contact with the modern market economy has involved the people in a semi-commercial production system. Because the projects have been Government enterprises the inhabitants may have felt little attachment to them, but they have developed an opportunist attitude and learned how to grasp any advantages offered to them. The people have acquired some experience of modern conditions.

The result has been, on the other hand, an increasing lack of confidence in the schemes as an acceptable model for development. But, as an unintended consequence, people have come both to lean on off-farm activities and to adhere more to their traditional mode of living by livestock keeping, making use of the natural resources in Central Butana and the Schemes of New Halfa, Rahad and the Mechanized Farming Schemes. The feeling of the people that these schemes are government property has made them "transient" users who rarely become involved. The people feel that such enterprises are beyond the community capabilities to attempt to reform, and that only the government can make the necessary improvements. The main advantage of contact and involvement in such projects has been to help people review their past herding strategies and reciprocal livestock-agriculture-livestock investments (Abu Sin, 1982). To adapt to such situations carefully worked out family divisions of labour have been devised which allow for participation of the people according to ability and perceived needs.

The market economy orientation pursued by development plans directed by foreign investors and World Bank policy has subjected the traditional sector to dramatic pressures of adjustment. The emergence of a "rural economic elite" is manifested in commercialization of livestock husbandry, agro-pastoralism, traders, middlemen etc. In spite of the large control of use of resources by this "elite" they represent only a small proportion of the population. This segment mostly come from outside Central Butana proper, yet has access to it under the communal tenure system and have political and financial power which can hardly be challenged by the poor majority. This has given the majority no option but to endeavour to consolidate their power through local community-based strategies. Extended family co-operatives which buy tractors, dig wells, rehabilitate services and control village pasture lots have emerged in most of the settlement clusters in Central Butana.

The top down process of development which dominates in Central Butana, and indeed all Sudan planning, exhibits basic assumptions which run counter to the effective planner/community based dialogue advocated by James Varies (1983) summarized below.

ENVIRONMENTAL SETTING

Central Butana constitutes a uniform ecological unit similar in climate, soil, terrain and resultant production systems and socio-economic structures. It has a simple uniform resource base, which differs only at the micro-scale eco-system (Abu Sin, 1970, 1975, 1982).

Table 1. *Traditional and dialogue approches*

Traditional Approach	Dialogue Approach
1. Planners assume an educational attitude. They teach the community to accept.	1. Both are involved in the learning process.
2. Planners as experts know everything and community knows little or nothing.	2. Both have knowledge to contribute to joint learning
3. Planners assume monopoly of knowledge	3. Nobody to assume that they have the monopoly of knowledge
4. Planners assume right to concern and people are to be concerned about.	4. People are encouraged to think independently and to get concerned about themselves.
5. Planners are more active in process of learning, decision-making and implementation.	5. Both are active in all phases at different levels.
6. Planners' perception is national in a simple balance of material gains, and people are assumed to grasp such an objective.	6. Compromise of people's and national concerns to attain objective of development and sustainability.

Source: Modified from Varies, 1983.

It covers an area of 35,000 km^2 (Lat. 13°30′ and 15°30′ N. and Long. 33°34′ and 34°30′ E.). Administratively it is confined to Butana Rural Council, the boundaries of which marked the ecological boundaries of Central Butana. The Shukriya, originally a nomadic tribe, have laid claim to all Butana for over 400 years. Settlement and production systems, although fairly evenly distributed in clusters around water sources, are almost identical. This similarity helps in replicating identical development ventures with the minimum cost and effort. Relief and drainage systems are the main control variables of land use, settlement and service distribution. The pattern of mobility and grazing use radiates from these clusters in approximately uniform radii in years of normal rains. Each sub-tribe or clan of the Shukriya tribe is the base for micro-scale development projects which match both the human and natural resource bases. At present the population distribution reflects the association between population size and clan micro-scale resource bases.

The family continues to be the main socio-economic unit within an integrated extended family and the clan. It is worth mentioning that the low locality attachment, assumed to be associated with pastoral nomadism, is now replaced by strong locality attachment as represented by the stability of the clustered settlements in the area.

Family division of labour is carefully observed under a form of collective decision-making. The Central Butana economy started as pure nomadic pastoralism and has passed through a gradual transformation to a mixed economy of livestock breeding and casual rainfed agriculture. At present we encounter a spectrum of production systems running from semi-nomadic pastoralism to agro-pastoralism. The systems only vary in degree of emphasis and are, in an interchangeable manner, dependant on the environmental conditions and cyclic changes. Livestock herd sizes are small, which indicates a state of equilibrium of carrying capacity and animal population under normal environmental conditions. Therefore, I believe that these conditions, plus the dissatisfaction that has resulted from top-down trials of development, has prepared Central Butana for community-based options of sustainable development.

The essentials for community participation and involvement in sustainable development options and how favourable conditions are in central Butana are summarized in Table 2.

Table 2. *Participation essentials and favourability conditions*

1. A closely knit institutional network	1. Customary family and tribal frameworks are functioning
2. Community organization to be based on cost-benefit perspectives	2. Increasing tendency of transformation of social linkages into economic and cost-benefit linkages.
3. Important role of leadership	3. Role of leadership at extended family or clan is taking economic shape in the form of family based enterprises.
4. Shrinkage of voluntary participation	4 Unpaid working for family is being replaced by working for income even within new family structured enterprises.
5. Division of labour and clear assignment of responsibility.	5. Clear identification of function in family-based enterprises.
6. Ability of leadership to mobilize the community.	6. Due to political changes institutional leadership limits its function, giving room to socially accepted institutions based on the family and clans.
7. Use of relevant technology and project ideas and execution procedures. 8. Ability to rank priorities and strategies.	7–8.Use of relevant and intermediate technology is limited, with relatively increasing ability to rank options on a purely economic basis.
9. Homogeneity of production system.	9. Production systems in all forms are well integrated and linked in a family balanced manner.
10. Desire to attain basic needs as a pre-condition for development.	10. Inhabitants are engaged in rehabilitation of basic services and infra-structure.
11. Narrowing the "perception gap" between planners and community.	11. Involvement of community representatives in decision-making in irrigated projects.

AREA DEVELOPMENT SCHEMES: A TRIAL OF COMMUNITY-BASED DEVELOPMENT

Central Butana was chosen by UNDP, as one of five areas in similar ecological settings, for an experimental attempt to introduce community-based development. The five areas are, El Obeid rural area, Umm Kaddada, Idd el Ghanam, Lower Atbara and Central Butana. The trial was based on:

1. participation of both planners and community in a joint model,
2. accommodation of local needs and strategies as identified by the community, to be tested against planners' beliefs, funds and available technology,
3. satisfying the basic needs and gradually increasing them in line with community and resource capabilities and people's willingness,
4. attaining a basic dialogue between planners and people to accommodate and blend the perception of planners with their own,
5. education throughout the community to ensure participation and economic returns.

To introduce and consolidate the model of community-based development in Central Butana, the UNDP adopted social animation. The aim was to test perception, willingness, participatory capabilities and institutions. Social animation in this case was pursued along the following lines and phases:

Phase I Training of social animators who were carefully selected for the task.

Phase II Testing community potentials for participatory abilities. This allowed them to identify their own problems and solutions. Participants are required to rank and prioritize problems and attempt to formulate relevant projects and how to execute them.

Phase III Dialogue phase, narrowing down priorities against the reality of the project having initial funding abilities.

Phase IV Scheduling, execution procedure, identification of implementation, details and logistics. Monitoring and follow-up steps were discussed with the community. This procedure is based on the following main assumptions.

Normally, a community responds positively to what they heard at the level of 20 per cent. In other words 20 per cent of community will respond and accept what they hear. About 40 per cent respond to what they hear and see. About 80 per cent respond positively to what they hear, see and do. This assumption proved to be valid in the social animation trial in Central Butana. Social animation was guided by the following:

a) Animators were instructed to acquire a good knowledge of the prevailing nomadic culture, social networks, institutions, beliefs, etc.
b) To share in the community's activities and respect the social order without being so affected that they loose their sense of mission. They should give each person the feeling that his/her role is vital and indispensable.
c) Animators are to avoid being bossy and knowing everything and try to avoid the government officials' attitude and behaviour, and be able to explain their mission in an easy and clear way.
d) They should not over-raise community hopes and should not promise what they can't give (Hamid Ali, 1990). The animators centre of action was sited in the area, unlike the case of the other four development areas. The advantages of siting the centre of action in the target area are:
    i. Increased contact with target groups and deeper understanding of the society.
    ii. It gives the community the confidence that the project is a reality and animators are serious and keen. It corrects the feeling that government officials are "delicate" and can't stay with a community.
    iii. Normalize relationships with project administration and lead to narrowing the "perception gap".

iv. Daily checking of participation acceptance message, i.e. level of keenness and willingness of the community. This has given social animators sound insights from which to formulate the relevant projects. As the result of a prolonged period of animation, the social animators arrived at a set of socially acceptable, technically and financially feasible projects.

These projects were also tested for environmental sustainability and are now ready for execution. These projects do not exhaust the development option, but are relevant to Central Butana at its present stage of environmental degradation.

CONCLUSION

Butana, like similar areas in Sudan, had been subjected to the classical top-down purely government initiated approach which had been targeting areas of high potential in the district. The community is expected to accept and adjust. Central Butana which constitutes the most marginal part of Grand Butana, had not received one large-scale project. Most of development attempts are the primary ones of services and simple infrastructure.

Moreover, under the communal resource tenure, by which any government project is a common property to be used by everybody, Central Butana has been subjected to serious degradation. To recover the best development option is a community-based attempt to match the resources and challenge the communal tenure system in order to attain an acceptable distribution of population and also attain sustainable growth and a better quality of life for all the people.

REFERENCES

Abu Sin M.E., 1970: *The Regional Geography of Butana, North of Railway*, MA Thesis, University of Khartoum.

Abu Sin M.E., 1975: *Population Mobility in North and Central Sudan*, PhD Thesis, University of London.

Abu Sin M.E., 1982: "A Change in Strategy of Animal Herding Among the Nomads of the Butana: Eastern Region" in G. Heinritz (ed), *Problems of Agricultural Development in the Sudan*, Gotlenger, F.R.G.

Abu Sin M.E., 1985: "Planners and Participants Perception of Development in the Semi-arid Lands of Sudan: A Case Study of Khashm El Girba Scheme" in H.R.J. Davies (ed), *Natural Resource and Rural Development in Arid Lands: Cases from Sudan*, UNU.

Awad M.M., 1964: "Sedentarization of Nomads in the Butana Region: North Sudan", *Bulletin de la Société de Geographie d'Egypte*, Vol. XXVII:15–20.

Hamed Ali, 1990: *Animation in Central Butana*, Typescript, (forthcoming).

Ibrahim M.A., 1989: *Environmental Degradation in Central Butana*, MA Thesis, Department of Geography, University of Khartoum.

Varies James, 1983: "Extension, Training and Dialogue: A New Approach". *Tanzania Journal of Adult Education*, Dar es Salaam.

Camel Milk Output: Evidence From Somalia

Mohamed Said Samantar

The Somali economy is mainly dependent on livestock, for about 40 per cent of the Gross Domestic Product comes from the livestock sector, and around 80 per cent of the foreign exchange is obtained from livestock export. Almost 60 per cent of the population is engaged in livestock rearing.

This is mainly due to the country's environmental and geographical conditions:

(a) The country's total surface area is 638,000 sq. km, out of which 45 per cent (i.e. 287,100 sq. km) is suitable for livestock raising.
(b) The climate, shared with the other tropical countries, with temperatures ranging from 25–35° Celsius, is suitable for raising livestock. The two rainy seasons, Gu'(April–June) and Deyr (October–November) give the nomads relief from the difficulties encountered in the dry seasons.

The main consumption product of a camel herding economy is not meat, but milk. Slaughtering of camels only takes place on rare occasions of great significance; a normally functioning subsistence pastoral economy principally produces milk.

Unfortunately information on the basic parameters of this important item have not been available to planners in Somalia before the mid-seventies.

PROBLEMS IN ESTIMATING CAMEL MILK OUTPUT

Conceptual and practical problems abound in estimating camel milk output. It is essential to understand the basic limitations to estimating camel milk data. The problem arises from the fact that we are dealing with a nomadic pastoral economy. One of the important facts is that there have been very few serious studies on the subject so far, and because most observers know little about the nomadic economy. It is difficult to estimate the camel population because of the seasonal migrations, the crossing of national boundaries and the herders' reluctance to give the exact size of their herds.

It is interesting to show how different estimators came up with different figures for camel milk output in 1978. In that year there were at least five different estimates. The first GDP document of national accounts in Somalia did not include an estimate of milk production. The second document of national accounts came up with a milk output figure of 451 million litres, not classified by type of livestock (Table 1). One can easily see the contradiction between these data and the data on camel milk output (Table 2). Most of the figures in the latter are actually much higher than those for total milk output as shown in Table 1.

Almost simultaneously with, or very soon after, the second document of national accounts the International Bank for Reconstruction and Development (IBRD) came up with an estimate of milk production. Although the production coefficients used were different from Jamal's (and incorrect), the methodology used was the same as Jamal's without any previous experience or previous studies to lean on. The estimate obtained was 1,066 million litres.

IBRD had also suggested a model to arrive at livestock population estimates at any given time, based on birth and death rates:

$$X_t = X_0 + X_0(B_0)(R_0) - X_0(M_0) - E_0 - C_0$$

X_t = herd size in the followin g year
X_0 = herd size in the current year
B_0 = percentage of breeding females in the herd in the current year (camels 45)
R_0 = lambing rate or calving rate in the current year (camels 35)
M_0 = mortality rate in the current year (8).
E_0 = livestock exports in the current year.
C_0 = domestic consumption in the current year.

These coefficients were applied in normal years; in the other years modifications were made to reflect the weather conditions. As far as concerns camel milk estimates the model's importance is twofold:

(i) it determines the milk supply using as coefficients the breeding rate and the calving rate, therefore the lactating rate;
(ii) it determines the change in camel numbers (stock change) directly.

As will be shown by the following figures, IBRD had erroneously applied the breeding coefficients to the total population, equating breeding with lactating. There was no question here of a "fit" with estimates from the consumption side: thus one can not help wondering whether, if IBRD had realized its mistake in using the production coefficients, it might not have chosen higher yield figures for milk (because its yield estimates are much on the low side) to square their production estimate with the consumption estimate.

IBRD tried to reconcile its production-side figures with those derivable from the consumption-side. A "fit" was obtained but unfortunately with incorrect figures. As shown below, very high lactation coefficients were used because of the error in confusing breeding females (b_0) with lactating females (r_0). Normally $b_0 \times l_0$ would give r_0. For camels the following figures were obtained:

Population	Breeding	Calving	Yield	Camel Milk Output
6.77 (m)	–45%	35 %	350 l.p.a.	(6.77x0.45x350)=1,066

The correct procedure should have been (6.77x0.16x350)= 379, the lactating rate being equal to (0.45x0.35) 0.16. Perhaps to compensate for the higher lactation coefficients, IBRD used rather low yield figures. Therefore the IBRD figures also have to be rejected as technically wrong. The methodology proposed by IBRD was applied by Mr. Cummings-Palmer (ECA, Economic Commission of Africa) in the next effort to revise Somalia's milk output. He used IBRD's coefficients for breeding and for yield. Cummings-Palmer corrected IBRD's mistake of equating breeding with lactating, although nowhere is the basis for this noted down for future researchers. The coefficients used were lower: so Cummings-Palmer gets a lower estimate of milk output, that is just 54 per cent of IBRD's estimates.

The next attempt at revising the national accounts was made in 1983 by Mrs. Nguyen—the author of Somalia's first national account. Unfortunately, no details are available in this report but, as far as one can make out, no changes were made as to milk estimates. In the meantime a team from the Swedish Institute of Statistics was assigned to the Central Statistical Department of the Ministry of National Planning, under the project "Development of National Accounts Statistics". They did not make any contribution to livestock figures.

The last revision of milk estimates was made by Tim Jones (ECA) in July–August 1985. The milk output figure he obtained was classified into camel, cattle, sheep and goat milk. His camel milk output estimate was 603 million litres. Tyler used a downward revised stock population (6.3–5.3) with coefficients of milking animals and milk yields, obtained from FAO expert J.A. Willby (FAO animal husbandry and range specialist, since deceased), to work out his own estimates of milk production. The figure he came up with was: (5.3x0.14 x1,200) 890. This figure was based on the same calculations that Jamal had used (6.3x0.25 x1,100 = 1,732.5) and, compared to the official estimate, it was too high. Tyler said Jamal's figure was not near the "truth". If that is the case, it is not clear what he thought of his own figure and what he intended to do with it. Tyler failed to understand Jamal's methodology, because the one he proposed in its pure form was identical to Jamal's. Tyler's and Jamal's coefficient differ greatly from the "official" ones. Tyler thought Jamal's coefficients of milk animals and milk yields were high, but he himself practically used the same coefficients for his macro estimation of milk production. Then he seemed to set aside this figure in favour of the national estimates, even though it was around one half of his own. The impression thus created was that he disbelieved his (and Jamal's, of course) estimates in favour of the data in the national accounts.

Jamal was writing at a time when the national accounts estimated milk production at 451 million litres (1978). Using the most probable coefficients, he arrived at a camel milk output of 1,732.5 million litres. As we are dealing only with camel milk we can accept Jamal's lactating coefficients, because he used a blanket rate of 25 per cent for all species as a sort of crude weighted average. The difference between Jamal's and Tyler's estimates arises from:

(i) the estimate of the total camel herd, and
(ii) the coefficients used for lactating camels.

Before Jamal completed his study, the government revised the total herd estimates downwards. Had he been working with this new figure, his production estimate would have been (5.3x0.25x1,100) 1,457.5 litres which is almost identical to the figure Tyler obtained using his own coefficients.

CAMEL POPULATION ESTIMATES

Among the major camel raising countries Somalia has the largest camel population, estimated at 5,298 million in 1975 (Census 1975), 6,601 million in 1987 and 6,680 million in 1988 (FAO Yearbook, 1988). Somalia possesses at least 35 per cent of the camels in the world. Moreover, the number of camels in the country is increasing at a tremendous rate as shown by the following figures:

1975 (5,298,000); 1987 (6,601,000); 1988 (6,680,000).

In the last two years camels have increased by 1.2 per cent (70,000 camels). If we calculate the growth rate over several years it appears to be roughly 1.5 per cent per annum. However, things are different in other camel raising countries; in some the number of camels has been decreasing while others maintain a constant rate.

LACTATING RATE

The lactating rate (proportion of lactating females) is defined as the percentage of females lactating out of the total herd; this should not be confused with the proportion of actually milked females, which is potentially lower: not all lactating females can always be milked for human consumption, for in the dry seasons, when the milk yield declines and/or at a late stage of the lactation cycle, the herders have to stop milking some females.

The breeding and calving rates are necessary to estimate the lactating rate in the total herd. It suffices simply to multiply the two. Here below are various estimates from different sources.

BREEDING RATE

The breeding rate is among the major coefficients in estimating camel milk output. The life cycle of the camel, starting from sexual maturity to the breeding stage, is worth considering. The breeding rate (or proportion of breeding females) is defined here as the percentage of mature females, i.e. old enough to become pregnant, out of the total herd.

However the age at which a camel is considered mature varies between reports. For instance, based on information collected from a sample of Somali camelmen (Hussein, M. Ali, 1984) puberty was estimated to occur at 5 years of age but sexual maturity was considered to occur only at 7 years of age when they were mated. From the discussion it emerged that the age of sexual activity for the females of the study herd was considered to be 4–5 years, while males are kept from breeding until age 6–7. According to data based on the Yaqbari-Weyne area study, it is claimed that Somali female camels reach maturity at about age 4 (Hjort and Hussein, 1986). Nevertheless the categories used by Hussein (1987b) are inconsistent with those in Watson (1986), Elmi (1989) and Boykin (1988). The variation could be explained by regional differences in the age structure of camel herds, which can be considerable (Watson, 1986), and in the more or less long recovery cycles of drought-affected areas.

The available data on the percentage of breeding females for Central and Southern Somalia are fairly consistent. This is shown in Table 5 based on the Yaq-Bariweyne area study. The proportion of breeding females should therefore be between 62 per cent and 43 per cent. That will roughly amount to an average of 50 per cent.

CALVING RATE

It is essential to note the interval between two successive breeding stages and the number of births a normal camel can possibly have. This division greatly simplifies the calculation of the camel population. The calving rate relates to the number of breeding females and is usually calculated by comparing the number of calves born in a certain period to the number of females exposed to males. As there seem to be few age-based calculations of camel fertility, it is taken for granted that the "calving rate" is calculated in relation to the total number of adult female camels, and defined as the percentage of she-camels giving birth during one year, out of the total number of grown up heifers and mature female camels. Regular biennial calving implies, of course, an annual calving rate of 50 per cent among fertile camels. It is virtually impossible to achieve a much higher calving rate even among ranched camels. Conception takes place in the rainy season. The female will then give birth during the next season, i.e. mating and calving coincide, but the female cannot be impregnated again until the next season, one year later.

MILK YIELD

The data on the milk producing capability of camels are not very accurate and only subjective estimates are available. Very scanty information is available on the amount of milk consumed by the calf and the yield estimates relate only to milk available for human consumption.

The officials of the Ministry of Livestock, Forestry and Range determine the camel milk yield in the following way:

(Daily Milk Yield x Calving Interval) – Amount left for the calf.

The resulting value is divided by three years. This method of dividing by three is erroneous from the conceptual point of view. The proper procedure should be to divide it by two years, i.e. the interval between two lactation periods. This period of two years is equal to the calving interval. The upper limit of the lactation period (18 months) instead of an average figure (8+18)/2 = 13 has been used (see Table 7).

CALVING INTERVAL

The calving interval is the period intervening between two successive births. It differs from country to country and from area to area. In some countries when a female camel gives birth it is allowed to mate again after six months, in other countries after nine months or even more than one year. The duration of pregnancy is 12 to 13 months: if this is added to the interval between birthing and mating, the calving interval is 24 or 25 months. As pregnancy advances, the milk yield decreases until the udder dries up. Since most of the Somali pastoralists live on camel milk, the camel herders keep the males away from the milking females soon after parturition.

LACTATION PERIOD

Although complete studies of the camel's lactation period have not been made, 18 months is the most commonly mentioned upper limit. Most camels give milk for a longer period of time than any other domestic animal. According to Leese (1927), as quoted by Williamson and Payne (1965), the dam will go dry after nine months if desert-fed, but will continue giving milk up to 18 months if well nourished. Leupold (1968) gives 9 to 18 months as normal for one-humped camels and 16 to 18 months for the two-humped variety. Mares (1954) gives 8 to 18 months as the range of the lactation period of the Somali camel.

There is a direct proportional correlation between the calving interval and the lactation period, i.e. the increase or decrease of either reflects on the other. In other words, a longer calving interval implies a long lactation period and vice versa. For example, if the mating of a female camel occurs again 12 months after parturition the lactation period is estimated to be 13–18 months; but if the calving interval is 21 months, the lactation period will be 10–15 months. This correlation has a limitation: if the interval exceeds two years it will no longer hold. Many writers and camel raisers believe that if the female camel ceases breeding it will go dry. As mentioned earlier calving is always mono-seasonal. When the average length of lactation drops below one year, scarcity of milk can be expected in the period just before the rains, when the new calves are to be born.

TREND OF NOMADIC PER CAPITA CAMEL MILK OUTPUT

On the basis of the 1986 census, the Ministry of National Planning and Juba Valley Development has officially announced Somalia's total population at around 9 million (April 1990). The annual population growth rate is 3.1 per cent. About 60 per cent, or 5.4 million people, follow a nomadic or semi-nomadic life-style. They rear mainly four species of livestock, namely camels, cattle, sheep and goats. The total milk output in 1989 was 2,317 million litres as estimated by MNP (National Accounts Aggregates 1977–1989). From these two sets of figures we can estimate the average annual per capita milk output per nomad, as well as the average annual per capita camel milk output. These averages are 502 and 194 litres respectively. If the daily per capita milk output is calculated for the 365 days of the year, the averages come to 1.38 and 0.53 litres respectively. This result is highly inconsistent with the per capita milk consumption even if only the 1981 figures are considered. In fact, the IBRD used the same consumption survey that Jamal had used in his 1981 study. It showed nomadic milk consumption of 4 litres during the wet season and 0.5 litre during the dry season. IBRD took the wet season to mean five months whereas in the original source it is given as March to October, i.e. eight months. Obviously, this is not a wet "season" in terms of rain but rather, one imagines, a measure of the time during which nomads could consume 4 litres of milk per day. The other issue, which stems from the above computation, is that the availability of per capita milk for the nomadic community presents a declining trend as is illustrated in Figures 1 and 2.

This is the kind of gloomy scenario with which the nomadic population is confronted. There is no doubt that, if different assumptions on milk output coefficients are considered, probably an opposite picture is obtained. As Jamal states: "... lots of arbitrary and *ad hoc* changes are made on the camel milk figures, to the extent that they lose all meaning. The problems with milk estimates in Somalia are that the production coefficients are not known. Thus we get a situation where the existing estimates differ from each other. Clearly, one's whole perception of the economy changes whether one believes in one extreme figure or the other." In any event, I personally recommend that the Nomadic Survey, envisaged for 1993 by IBRO-ADB in collaboration with the Somali government, should design a well-structured questionnaire in such a way as to solve at least the contradictions between the figures concerning the nomadic population and milk production.

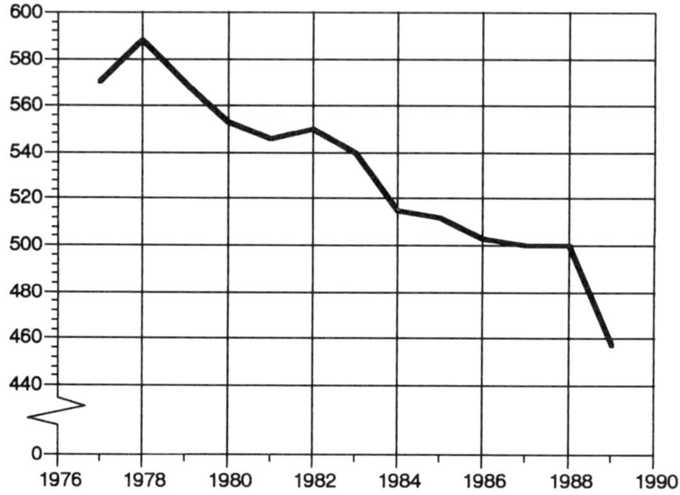

Figure 1. *Nomadic milk per capita*

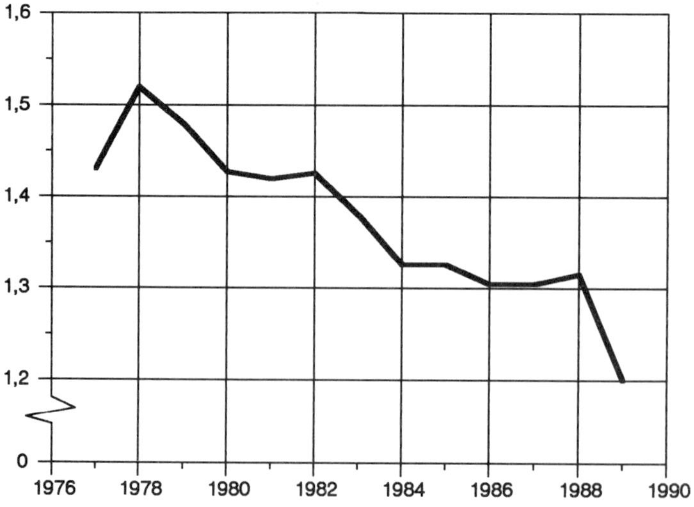

Figure 2. *Nomadic camel milk per capita*

Table 1. *Estimates of milk output in 1978*

Date Produced	Milk output (ml)
Nat. Acc. Dec -79	451
IBRD April -80	1,922
FAO	745
ECA (C-P) Aug -81	1,033
Nat. Acc. -85	1,309
Nat. Acc. Aggr. 87/88	2,061

Source: Vali Jamal, Somalia: Economics for an Unconventional Economy

Table 2. *Camel Milk output in various estimates, 1978* (in million litres)

IBRD	379 (1,066)
Nat. Acc. -85 (ECA)	479 (603)
FAO	199 (198)
Tyler/Willby	890
Jamal	1,711 (1,560)

Source: Vali Jamal, Somalia: Economics for an Unconventional Economy

Table 3. *Various estimates of camel population in Somalia, 1978*

	Population in millions
IBRD	6.77
Nat. Acc., current	5.70
FAO	5.40
Tyler/Willby	5.30
Jamal	6.25

Source: ILO/JASPA, 1989

Table 4. *Lactation rate*

IBRD	45
Nat. Acc. -85	14
FAO	6.7
Tyler/Willby	14
Jamal	25
Plan Task Force	40
MNP proposal	14.2
Dahl & Hjort	17

Source: ILO/JASPA, 1989

Table 5. *Breeding rates of camels in central and southern Somalia*

Region	Total	Breeding females	Breeding rate as %	Data source
Central	76	47	62	Elmi (1989)
Southern	83	36	43	Watson (1986)
Somalia	--	--	45	IBRD (1980)

Source: Urs Herren, 1989, ILO/JASPA, 1989

Table 6. *Yield in litres per camel*

IBRD	350
Nat. Acc. -85	600
FAO	550
Tyler/Willby	1,200
Jamal	1,095
Plan Task Force	270
MNP proposal	500

Source: ILO/JASPA, 1989

Table 7. *Lactation periods*

8–18 months	Mares (1954)
9–18 months	Leupold (1968)
9–18 months	Leese (1927)

Source: Dahl, G. and Hjort, A., 1976

Table 8. *Trend of milk output, nomadic population, nomadic milk per capita, (1977–1989)*

Year	Total milk in million litres	Total population in millions	Nomadic population	Milk per capita in litre
1977	1,922	6.1	3.66	525
1978	2,061	6.2	3.72	554
1979	2,122	6.5	3.90	544
1980	2,102	6.7	4.02	522
1981	2,141	6.9	4.14	517
1982	2,218	7.1	4.26	521
1983	2,218	7.3	4.38	506
1984	2,205	7.6	4.56	484
1985	2,264	7.8	4.68	484
1986	2,315	8.1	4.86	476
1987	2,374	8.3	4.98	477
1988	2,478	8.6	5.16	480
1989	2,317	8.8	5.28	439

Source: Nat. Accounts Aggregates, 1988–1989

Table 9. *Trend of camel milk output, nomadic population and nomad camel milk per capita (1977–1989)*

Year	Total camel milk output (Million liters)	Nomadic population	Nomadic camel milk per capita in litres
1977	766	3.66	209
1978	798	3.72	215
1979	816	3.90	209
1980	814	4.02	203
1981	827	4.14	200
1982	858	4.26	201
1983	866	4.38	198
1984	861	4.56	189
1985	880	4.68	188
1986	895	4.86	184
1987	909	4.98	183
1988	941	5.16	182
1989	881	5.28	167

Source: Nat. Accounts Aggregates, 1988–1989

REFERENCES

Boykin Calvin C. , 1988: *Benefit-Cost Analysis of Range Improvement Interventions.* Mogadishu, Somali Central Rangelands Project and Louis Berger International.

Dahl Gudrun and Anders Hjort, 1976: *Having Herds—Pastoral Herd Growth and Household Economy,* Department of Anthropology, University of Stockholm.

Elmi Ahmed A.,1989: "Camel Husbandry and Management by Ceeldheer Pastoralists in Central Somalia". *ODI Pastoral Development Network Paper* 27d, the Overseas Development Institute. London.

Herren Urs, 1990: "Milk Production and Marketing in Yaq Bariweyne area, Southern Somalia", *Camel Forum Working Paper* No. 32.

Hjort Anders and Mohamed Ali Hussein, 1986: "Camel Herd Dynamics in Southern Somalia: Long Term Development and Milk Production Implications", *Camel Forum Working Paper* No. 13.

Hussein Mohamed Ali (ed), 1984: "Camel Pastoralism in Somalia" Proceedings from a workshop held in Baydhabo April 8–13. *Camel Forum Working Paper* No. 7.

Hussein Mohamed Ali, 1987(a): "Emic Notes on Camel Breeds in Somalia". *Camel Forum Working Paper* No. 17.

Hussein Mohamed Ali, 1987(b): "Traditional Practices of Camel Husbandry and Management in Somalia", *Camel Forum Working Paper* No. 19.

ILO/JASPA, 1989: "Generating Employment and Incomes in Somalia", Interdisciplinary Employment and Project-identification Mission to Somalia, Addis Ababa.

Jamal Vali, 1987: SOMALIA: Economics for an Unconventional economy, *mimeo.*

Leese A.S., 1927: *A treatise on the one-humped camel in health and in disease,* Hagues and Son. Stanford.

Leupold J., 1968: "The camel. An important domestic animal of the sub-tropics" *Blue Book Vet Prof,* Inst. Trop. Vet. Med., Justus-Liebig University, Giessen, pp. 1–6.

Mares R.C., 1954: "Animal Husbandry, animal industry and animal disease in the Somaliland Protectorate —Parts I & II", *The British Veterinary Journal,* vol 110, Nos. X and XI.

Ministry of National Planning, 1989, *Somalia in Figures,* Central Statistics Department, l0th edition, Mogadishu, Somalia.

Taneja G.C., 1980: *Present Position and Scope of Improvement in Camel Production in SDR,* Rome, FAO.

Tyler G.J., 1983: *Somalia: A Case Study on Rural Poverty,* Rome, FAO.

Watson R.M., 1986: "A Programmetric Sampling of Camels to deduce the age structure of the populations in Southern Somalia", *Camel Forum Working Paper* No. 12.

Williamson G. and W.J.A. Payne, 1978: *An Introduction to Animal Husbandry in the Tropics,* London, Longmans.

Ecological and Pastoral Production in Nigeria: Issues and Insights

Jerome O. Gefu

> Since I first grazed my father's animals on this reserve some thirty-five years ago, I have seen less and less of the local feed stuffs that we have become so used to and have valued so dearly. Many of the grass and legume species found on the reserve now are not as nutritious as those that were originally present. I have to send part of my herd away so as to reduce grazing and watering pressure. (Fulani herder, 1990)
>
> I have fewer animals than my father yet there was more forage available for those animals than there is now. Maybe a time will come when we shall no longer be able to raise animals if there are no grasses to graze and no browse plants. (Fulani herder, 1990)

INTRODUCTION

The above are the words of a pastoralist who has settled permanently on one of the grazing reserves in Nigeria. His concern, like that of many other pastoral producers, centres round pasture and water, the essentials for livestock production. These typical statements illustrate the crisis of pastoralism in African arid and semi-arid lands, in which the restrictions imposed on the mobility of pastoral herds have contributed to environmental degradation and increased both competition and pressure on the ecology. The concerns and apprehensions of the Nigerian pastoralist cited above recall the warning observation that rangelands in the American West now, "produce less than half the biomass they did before being damaged by the grazing industry" (Jacobs, 1988:15). The combined activities of livestock grazing and other activities of man and nature have contributed to transform the pastoral landscape from relatively lush vegetation to scraggy stands of less desirable plants and, in some cases, just bare land. Lower production of milk and meat leads in turn to an increased dependence on grain and greater need for fuelwood for food preparation (Dahl, 1985).

More than 60 per cent of tropical Africa is made up of arid and semi-arid areas, although less than 30 per cent of Nigeria's land can be classified as arid and/or semi-arid. Such areas support over 35 per cent of the Sahel population of about 116 million inhabitants (Kotschi, *et al.*, 1986).

Pastoral societies have, for several decades, constituted the focal point of research and development activities for various agencies. The amount of attention actually given to the people has varied, but development-oriented intentions have aimed at bringing some meaningful change in the well-being of the people and at the same time protecting the environment. But, in an attempt to create an environment conducive to production, those agencies and governments which have been concerned with preserving the environment, especially in pastoral areas, have, in fact, declared war on the environment with a tactical system called "range management" which utilized weapons of "range improvements". Ironically, it has been through efforts and attempts made to contain some of the perceived problems which threaten the environment and the well-being of pastoral societies, that ecological degradation has continued to manifest itself. Pastoral development projects in arid and semi-arid lands of the world have, in practice, aggravated the problem. The issues, therefore, are the proper identification of the real culprits and the damage which they have caused. We must, of course, recognize at the same time that the problems of ecological degradation are compounded where many range degradation agents come to play in the same ecosystem. So, to conceptualize the problem, there is a need to determine the levels of guilt both of grazing animals *per se* and the range management strategies which are being pushed in pastoral areas, and the effect of non-pastoral and other livestock activities on the environment.

Doubt has been expressed in many quarters in recent times about the classical claim that the settlement of pastoralists and intensive ranching are sound measures for conserving soil and securing a proper development of pastoral production. For example Enghoff (1990) writes; "man-induced soil erosion is more commonly associated with areas occupied by settled agro-pastoralists and agriculturalists than with areas under extensive production. Mobility of pastoral herds are, therefore, a prerequisite for efficient production in most pastoral areas". Due to the patchy and sporadic nature of rainfall in arid and semi-arid lands, high mobility of herds is often widely adopted as a production strategy. Mobile livestock production has the advantage of being able to graze, for a longer period, relatively young grass with higher nutritional value. The effects of long distance movements on animal production are often regarded as negative. However, if driving is done carefully, animals can even gain weight en route (Reusse, 1982; Skoupy, 1988). Pastoral projects often have as their long term objective the development of a land use system which will ameliorate or eliminate rangeland degradation as well as meeting the needs of the ever-growing population of sedentarized producers. However, sedentarization projects have often been accompanied by desertification, especially around watering points and other infrastructural amenity sites (Skoupy, 1988). Nevertheless, renewed efforts are being made, especially by aid agencies, to pursue sedentarization programmes. The Nigerian case is certainly not atypical of pastoral and livestock development programmes aimed to sedentarize pastoral producers. Arguments in support of and against sedentarization have been presented elsewhere (Gefu, 1987, 1988). Suffice it to remark here that the grazing reserve strategy and settlement of pastoral producers in Nigeria calls for renewed critical appraisal, in the light of some of the serious ecological stress being experienced in many project sites in different parts of the country. Ecological degradation is a function of several interacting variables so its resolution will require multidisciplinary efforts.

Most pastoral communities are found in fragile, sensitive and marginal lands where material production and the reproduction of the pastoralists way of life hinge on two important resources: viz forage and water. In many arid and semi-arid lands productivity of pastoral rangelands has continued to decline as the vegetation has been denuded. Vegetational zones, especially in the pastoral regions, have shifted southwards, and continue to do so, as a result of a combination of wood cutting and accelerated soil erosion. "Desert patches" extend continuously through overgrazing and wood cutting until they coalesce into local or regional deserts (Hellden, 1985).

LOCATING THE PROBLEM

In some cases the activities of man, consciously or otherwise, facilitate the ravaging action of nature. But can we blame ecological degradation in pastoral areas entirely on pastoral activities? Certainly not! Other human and natural agents contribute, such as population pressure, charcoal burning, fencing and other domestic needs for wood, road constructions and large-scale agricultural projects. Natural forces include low seasonal and annual rainfall and subsequent poor vegetation cover. Poor and/or sporadic rainfall adversely affect the availability of vegetation cover which makes the action of erosion (wind and/or run-offs) more pervasive and more devastating. Fuelwood extraction, for example, has become a very prominent activity particularly in rural Nigeria. Indeed, Africa has been observed to top the list for fuelwood consumption per capita. The contribution of fuelwood extraction to the degradation of the environment is associated with, among other things, rapid population growth; but, on the plus side, it has increased the availability of alternative cheap energy sources such as kerosene and encouraged policies of reforestation (Okaiyeto, 1985). There is an urgent need for us to learn how to reverse environmental degradation.

PASTORAL VS LIVESTOCK DEVELOPMENT

More often than not, pastoral development has been confused with livestock development programmes and policies. Although both concepts have overlapping notions and common referents, they can be distinguished on the basis of the target beneficiary of the policy or programme. For the sake of clarity I will state what I mean by pastoral development in contrast to livestock development because, as in so many areas of development, pastoral development has been used to mean different things in different socio-political contexts.

Pastoral development can be conceived as a mainly social development activity, aimed at the improvement of the standards of living of pastoralists through the provision of health, education, veterinary, water and other services together with institutional building for better systems of range management. Livestock development, on the other hand, is an economic activity based on cost recovery with the aim of achieving some or all of the following:

a. technical change and the introduction of new inputs such as medicine, vaccines, fodder farming and the use of engine power for watering,
b. specialization in livestock products through the adoption of modern marketing of dairy products and wool, hides and skins,
c. increasing the producers' ability to evade seasonality by transforming milk and meat into storable forms, usually through technical advancement and specialization,
d. development of integrated market outlets responsive to national and international demands (Mohamed Salih, 1990a and 1990b). In Nigeria the thrust of development has largely been livestock production. Concern about the perceived hardships of millions of Nigerian transhumant and landless pastoralists in the face of diminishing natural rangelands has caused both pastoral and livestock development interventions (often led by the Nigerian state) to be launched.

The emphasis on intervention on livestock development, as against pastoral development, can be seen as a major reason for the slight success achieved in the overall attempts to bring about concrete changes in the livestock production sub-sector. It was as a livestock development effort that the grazing reserves were initiated by the World Bank survey team in 1954. Grazing reserves were conceived in order to attract pastoral producers, especially the long distance transhumants, to settle permanently and become agro-pastoralists.

The stated and unstated reasons for sedentarizing pastoralists are mixed. They arise from a desire by the interventionist agency to bring prosperity, social and economic infrastructure to pastoralists on the one hand, and a covert intention to exert firmer political control over pastoral producers, and thereby integrate them into the mainstream of the economy (Gefu and Gilles, 1990) on the other.

The impact of grazing reserves in influencing, let alone improving, pastoralists' extensive production methods has not been impressive. Where grazing reserves have been established severe range degradation problems have often followed and pastoral production systems have suffered stress. In many instances overstocking, overgrazing and ecological degradation have been aggravated by interventions which were meant to improve the lots of the producers, as well as promote a system of prudent range land use for sustainable livestock production. The establishment of grazing reserves in Nigeria has not achieved the intended objective of attracting transhumant pastoralists to take up sedentarized life. Instead, it is those already sedentarized pastoral producers, found around the existing fringes of the reserves, that relocate to utilize the advantages that these reserves offer. A factor that has limited the impact of grazing reserves in Nigeria has been the difficulty in acquiring and gazetting patches of land to be designated as grazing reserves. The land acquisition issue has often led to a variety of inter- and intra-community conflicts between pastoralists and agriculturalists. Yet, access to and unhindered utilization of grazing land is at the centre of the grazing reserve programme. Where government acquired land is available the demand to use it becomes so great that ecological, social and economic problems arise.

According to a recent policy statement of the Federal Government, Nigeria is making available a minimum of ten per cent of the country's land for utilization as grazing reserves by pastoralists. But, as with so many previous official pronouncements concerning pastoralism and pastoral production, very little has actually been accomplished beyond the identification and demarcation of proposed grazing reserves (Suleiman, 1989). The problem of acquisition still persists. Of the 2.3 million hectares earmarked for the development of about 300 proposed grazing reserves in the eleven northern states, only 0.5 million hectares have been gazetted. The difficulty faced in the acquisition and development of grazing reserves, the corner stone of the state's policy, continues to mar pastoral development. So it is not surprising that grazing reserve projects have achieved very little, especially in the areas of effective range management and range improvement. Not only did these projects fail as a result of the unavailability of adequately gazetted lands but, even where grazing reserves have been established, numerous problems have surfaced which had not been anticipated nor addressed by the interventionist state. These problems are briefly highlighted in this short presentation.

PASTORAL DEVELOPMENT PHILOSOPHY

Sedentarization within the grazing reserve enclaves was vigorously pursued by the Federal Government with the strong backing of aid agencies, notably the World Bank. The ultimate goal was an aggregate improvement in the productivity of the national herd while, at the same time, enhancing the well-being of pastoral producers. This was to be achieved through the adoption of the improved livestock production techniques as practiced by ranchers in the West. It meant that proper range management procedures would have to be followed. This required destocking to reduce perceived overstocking, establishment of fodder banks, land holding on an individual basis, and so forth. These are situations and conditions which are alien to the pastoralist but which development agencies have continued to push.

Whether or not pastoralists subscribed to the acclaimed virtues of "improved" production methods was hardly asked, nor were the short- and long-term consequences of sedentarization considered. For instance, it is not to be expected that more permanent settlement will take place during good seasons, because there is no need, and in bad years pastoralists tend to emigrate to better pasture away from the grazing reserve enclaves. Moreover, the grazing reserves which have been delimited can rarely be adequate for the ten million or more pastoralists in Nigeria. Further, destocking has not been known to work in the past in pastoral schemes in Africa. The reasons are simple. Herds are very difficult to reconstitute after major natural disasters. It is difficult to restock, especially with larger ruminants, after a period of drought or other calamity. Producers in drought prone areas can receive only 60–70 per cent of the "normal" market value of their animals when they are forced to destock. At the end of a drought it is almost impossible to buy breeding stock even at inflated prices, because everyone else is trying to rebuild their herds and are, therefore, unwilling to sell off their heifers. The fluctuation between under- and over-stocking that usually occurs, as a result of the seasonal and annual variation in rainfall, has made both the concepts of destocking and of carrying capacity very fluid. It is hard to push a programme or a philosophy of safe stocking rates.

In most cases pastoralists accommodated within grazing reserves are known to exceed the stipulated carrying capacity. This must be expected. This means that there has to be a workable and efficient system of emergency accelerated restocking available, if destocking policies are to be accepted. Indeed, the concept of sedentarization, within the present ideals and claims of the grazing reserves, contain inherent contradictions. These, though unintended, always become manifest as a sedentarization programme progresses. For example, introduced grazing systems are increasingly being modelled, on one hand, after those of the West but certain attributes of the "traditional" Western pastoral systems are also castigated by experts on the other hand. This contradiction indicates that while seasonal movement is technically efficient (for example, there is considerable inter-regional movement of livestock and persons especially during severe calamities in Australia) the development trend is still towards sedentarization. Political and economic pressures are towards increasing formal tenure and subdivision which make movement more difficult. (See, for example, de Leeuw, 1990; Abel and Blaikie, 1990.)

Furthermore, the pattern of land allocation proposed to pastoralists of only ten hectares of land, (of which two ha to be put under crop cultivation, four ha for forage legume production and the remainder to be kept as "forage fallow") on an individual lease holding basis is not in line with the community spirit of most pastoral communities. The philosophy of individual land holding, it is said, forestalls the consequences forecast by the tragedy of the common thesis; the thinking of which is that a patch of land held in common would be mismanaged, but, if allocated to individuals, the owners would bear the costs of overstocking and overgrazing and therefore not do so. The forecasts of "the tragedy of the commons" thesis do not appear to hold for most pastoralists in Nigeria, who have satisfactorily managed their livestock on common pasture for generations.

Indeed, not only is the ten ha land allocated to each pastoral household grossly inadequate (considering the average herd size of a sedentarized pastoralist *vis-à-vis* the ideal carrying capacity), but it creates a series of problems, such as land fragmentation and consequent litigation. Related are the problems of shrinking cattle corridors and the scrabble for a wet grazing spot (*fadama* land), which can lead to inter-community or interethnic conflicts.

WHAT IS TO BE DONE?

If ecological stress and crises are results of various factors, it follows that the solutions must necessarily be broad based. In the first instance appropriate and relevant livestock and range management systems need to be carefully worked out. Such systems should be location specific, to take cognisance of the prevailing local ecology and the needs of the local people. Improvements in such things as controlled extension of water points, livestock population control adjustment and carrying capacity improvement, pastoral grazing strategy development, planting of fodder trees of economic importance and renewed aforestation programme are some of the actions that can have both short and long-term effect(s) in reversing rangeland degradation.

The provision of essential infrastructural amenities by government has, however, created its own ecological problems. For instance, the provision of watering points on different grazing reserves has attracted more stock and led to the concentration of herds around such limited water points. This situation has facilitated, in no small measure, the incidence of localized desertification or serious range degradation. Furthermore, where watering points exist, there is an attendant problem of frequent breakdown of the diesel-powered boreholes due either to inavailability of spare parts or non supply of diesel to run the engine. These are indications of little planning and/or coordination of water use. It should be acknowledged that water development is an important ingredient in livestock and pastoral development. It can be used as a strong means of opening up many grazing areas which have been underutilized in the past, but it can be a source of disaster if ill-planned. In the first instance, watering points should be located far apart; usually about 20–30 sq. km in between watering points should be maintained, depending on vegetation and canopy characteristics.

Furthermore, the existing motor-pumped boreholes should be replaced with manual devices. This has the advantage of serving a limited number of herds. The use of hand-driven pumps limits the number of animals to the capacity of the boreholes, as well as providing a more regular and reliable source of water as the incidence of break downs is drastically reduced.

The watering points should be monitored very closely to prevent serious range degradation around them. Each of the watering points should be closed for re-seeding and natural regeneration of vegetation for certain periods of the year, while alternative watering points are made available to pastoralists. To further prudent land use the sustainable carrying capacity of different rangelands should be determined while livestock and human populations are regulated in line with the determined carrying capacity. Where necessary policing methods should be utilized to enforce this regulation.

To complement the proper planning and utilization of rangelands, renewed efforts should be made to encourage inhabitants of arid and semi-arid areas to plant trees of economic value. Tree planting does not only prevent degradation of the soil, it also provides feed resources for livestock, food for human consumption and even cash. Trees meant for planting should have been proven to be well adapted to the ecology, be drought resistant and to be acceptable to the local people. The ongoing tree planting campaign of the Federal Ministry of Agriculture and Natural Resources is a step in the right direction. However, more could be achieved if more extensive contacts were made with the more remote parts of the drought-prone areas of the Sahel. Furthermore, extension and project staff should be better trained in afforestation matters in order to be more effective change agents. As an incentive to planting more trees for the protection of the environment from the advancing desert, a reward system should be introduced for the "best environmental protectionist" in a given zone, and attractive consolation prizes distributed to deserving tree planting participants.

For a lasting impact to be made, pastoral-related projects and programmes should be reasonably long-term. The present practice by which projects are supported by donor agencies for only a few years, after which they are left to the home government to manage, needs to be reviewed. Long-term pastoral projects of assistance (technical, manpower, financial, etc.) should replace the rather short-term ones. A minimum project phase of fifteen years should be considered before assistance is completely withdrawn. During this period concerted efforts should be made by both the host country and the aid organization to transfer honestly the necessary technical-know-how to local personnel, so that, at the time of withdrawal, a local crop of personnel have been developed and adequately trained to take over every aspect of the project effectively. Project components should include as many locally sourced ingredients as possible. This would greatly reduce post-assistance dependence.

In order to ease the stress on the present conventional grazing lands, new grazing areas should be opened up in the sub-humid zones where pasture is more readily available for year-round grazing. The argument that the sub-humid zone cannot effectively support large ruminant production because of trypanosomiasis can no longer be sustained. The presence of pastoralists with large herds of Bunaji (White Fulani) cattle in many parts of the sub-humid zone, and to some extent in the rain forest areas of south-eastern Nigeria, attest to the fact that the southern limit of the "cattle belt" has shifted extensively southwards. The healthy looking and well conformed Bunaji cattle found in the rain forests of south-eastern Nigeria give one an impression that the animals can develop immunity to trypanosomiasis, while utilizing the abundant forage for rapid growth and higher off-take rate.

As a measure to safeguard livestock and the environment from the ravages of epidemic outbreaks, drought and other natural disasters, early warning device systems should be worked out to alert the public. The Department of Pest Control Services should be better equipped to be able to combat livestock diseases as well as mount effective campaigns for the prevention of livestock disease.

The extent of ecological degradation caused by pastoral and livestock activities is not known. Nevertheless, environmental degradation has often been blamed on the extensive system of livestock management. There is a need to undertake a nationwide land resource survey in order to be able to document the extent and pattern of rangeland utilization and degradation. From the results obtained adequate land use planning and water development can be embarked upon.

Furthermore, since there are strong indications of high fuelwood demand especially by ruralites (who constitute about 80 per cent of Nigeria's population), and the growing demand for fuelwood and charcoal in urban and pre-urban areas, there is an urgent need to reduce the rate of wood extraction. This implies that alternative sources of energy to fuelwood and charcoal will have to be explored; the abundant by-products generated by the refineries around the country can provide cheap and ready sources of energy. With more families utilizing energy sources other than fuelwood and charcoal, less trees would be extracted, thereby reducing ecological degradation. Adequate safety measures should be worked out to reduce the incidence of home accidents through the use of the alternative home appliances.

More important, our approach to the study of ecological stress should assume a multidisciplinary posture. Demographic issues are as important as ecological, sociological and political ones. The need for a systems approach to facilitate a better understanding of the problems cannot be overemphasized. In this wise, participatory management of the natural vegetation, in which the general population is mobilized through identified pragmatic and effective opinion leaders, has to be adopted.

REFERENCES

Abel N.O. J. and P.M. Blaikie, 1990: "Land degradation, stocking rates and conservation policies in the communal rangelands of Botswana and Zimbabwe", *Pastoral Development Network*. Paper 29a. Overseas Development Institute, London.

Dahl G., 1985: "On Anthropology and Ecology: The prospects of sharing more than common shortcomings". In A. Hjort (ed) *Land Management and Survival*. The Scandinavian Institute of African Studies, Uppsala, Sweden, 63–69.

Derrick J., 1984: "West Africa's worst year of famine", *African Affairs*, 83. No. 332, 281–300.

Enghoff Martin, 1990: "Wildlife conservation. Ecological Strategies and Pastoral Communities: A Contribution to the Understanding of Parks and People in East Africa". Paper presented to the Workshop on Pastoralism and the State in African Arid lands. The Scandinavian Institute of African Studies, Uppsala, Sweden and the Department of Social Anthropology, University of Gothenburg.

Federal Office of Statistics 1986: National Integrated Survey of Households. General Household Survey. Lagos.

Gefu J.O., 1987: "Land use in Nomadic Pastoralism and the Issue of Sedentarization". In M. Mortimore, E.A. Olofin, R.A. Cline-Cole and A. Abdulkadir (eds) *Perspectives on land Administration and Development in Northern Nigeria*, Bayero University, Kano. Chapter 8:72–81.

Gefu J.O., 1988: "Livestock development paradox in Nigeria: Insights from some development perspectives". In L.A. Ega and T.K. Atala (eds), *Rural Development Policies in Nigeria*, Proceedings of the 4th National Conference of the Nigerian Rural Sociological Association, Ahmadu Bello University, Zaria. 72–93.

Gefu J.O. and J.L. Gilles, 1990: "Pastoralists, ranchers and the state in Nigeria and North America: A comparative analysis". Paper presented to the Workshop on Pastoralism and the State in African Arid Lands. The Scandinavian Institute of African Studies, Uppsala, Sweden and the Department of Social Anthropology, University of Gothenburg.

Hardin G., 1968: "The tragedy of the Common" *Science*, 162, 1243–1248.

Hellden O., 1985: "Land Degradation and Land Productivity Monitoring: Needs for an Integrated Approach". In A. Hjort (ed) *Land Management and Survival*. The Scandinavian Institute of African Studies, Uppsala, Sweden. 77–87.

Hjort A., 1985: "Interdisciplinary views on land management and survival: An introduction". In A. Hjort (ed) *Land Management and Survival*. The Scandinavian Institute of African Studies, Uppsala, Sweden. 9–15.

Hjort A. and M.A. Mohamed Salih, (eds), 1989: *Ecology and Politics: Environmental Stress and Security in Africa*, the Scandinavian Institute of African Studies, Uppsala.

Jacobs L., 1988: "Amazing graze: How the livestock industry is ruining the American West", *Desertification Control Bulletin: A Bulletin of World Events in the Control of Deserts, Restoration of Degraded Lands and Reforestation*, No. 17:13–17, United Nations Environment Programme.

Kotschi J., R. Adelhelm, W. Bayer, G. von Buau, J. Hass and A. Waters-Bayer, 1986: *Towards Control of Desertification in African Drylands*, The Commission of the European Communities, Brussels; Bundesministerium für wirtschaftliche Zusammenarbeit, Bonn; Deutsche Gesellschaft für Technische Zusammenarbeit, Eschborn, FRG.

de Leeuw P.N. and J.C. Tothill, 1990: "The concept of rangeland carrying capacity in sub-Saharan Africa: Myth or reality", *Pastoral Development Network* Paper 29b, Overseas Development Institute. London.

Mohamed Salih M.A. 1990a: "Pastoralism and the state in African arid lands: An overview." Paper presented to the Workshop on Pastoralism and the State in African arid lands. The Scandinavian Institute of African Studies, Uppsala, Sweden and the Department of Social Anthropology, University of Gothenburg.

Mohamed Salih M.A., 1990b: "Ecological stress and political coercion in the Sudan", *Disasters*, Issue 14.2.

Nuru S., 1989: "Research and development in pastoral production system in Nigeria". In Gefu, *et al.* (eds), *Pastoralism in Nigeria: Past, Present and Future*, Proceedings of the National Conference on Pastoralism in Nigeria. National Animal Production Research Institute, Ahmadu Bello University, Zaria, Nigeria. 52–62.

Okaiyeto P.O., 1985: "Desertification of Nigeria's land: Is Pastoralism or Fuelwood Extraction the Culprit?" Paper presented at the National Workshop on Ecological Disaster: Drought and Desertification. Organized by the Federal Ministry of Science and Technology, Held at Central Hotel, Kano. December 9–12.

Reusse E., 1982: "Somali's Nomadic Livestock Economy: Its response to profitable export opportunity", *World Animal Review*, 43, 2–11.

Skoupy I.J., 1988: "Developing rangeland resources in African drylands", *Desertification Control Bulletin: A Bulletin of World Events in the Control of Deserts, Restoration of Degraded Lands and Reforestation* No. 17, 30–36, United Nations Environment Programme.

Suleiman H., 1989: "Policy issues in pastoral development in Nigeria". In Gefu, *et al.* (eds) *Pastoralism in Nigeria: Past, Present and Future.* Proceedings of the National Conference on Pastoralism in Nigeria. National Animal Production Research Institute, Ahmadu Bello University, Zaria, Nigeria. 38–51.

Vainio-Mattila A., 1985: "Socio-economic Studies on Fuelwood Production in the Bura settlement scheme in Eastern Kenya". In A. Hjort, (ed) *Land Management and Survival.* The Scandinavian Institute of African Studies, Uppsala, Sweden. 128–137.

Watts M., 1983: *Silent Violence, Food, Famine and Peasantry in Northern Nigeria,* California University Press. Berkeley.

"Big men" and Cattle Licks in Oromoland*

P.T.W. Baxter

This essay is an attempt to set my understandings of an apparently isolated and very matter-of-fact piece of indigenous knowledge in its ethnographic and developmental contexts. That I have come to think that there is anything in the data which requires "understanding" follows from the many discursive conversations I have had, over the years, with David Brokensha which have directed my attention to the minutiae of daily life and of indigenous knowledge. These chats have been to me as enjoyable as salt is to Oromo livestock.

In the early nineteen fifties I carried out fieldwork among the Boran, one of the Oromo-speaking peoples of the Horn; I spent most of my time in Marsabit District in Northern Kenya. At that time the Boran were prosperous, self-sufficient pastoralists who subsisted almost entirely on the milk, blood, and meat which their herds supplied; they sold sheep and goats (and occasionally a bullock) to pay poll tax and fines levied for breaches of grazing regulations; to purchase cloth, blankets and household items such as enamel basins, mugs and tumblers, and supplementary foodstuffs such as sugar, tea, salt, and maize meal to vary their diet, and tobacco and coffee beans for their spiritual and social comfort. The pastoral life was arduous, but the people were largely self-sufficient and were still, or so it seemed, users of the encroaching market economy rather than its victims. A few younger sons joined the police or went to work as stockmen on European farms down country, but that was as much for the experience as for the wages. Boran recent sufferings and destitution make that period seem like a Golden Age.

---

* This paper originally appeared in *Social Change and Applied Anthropology: Essays in Honour of David Brokensha* edited by Miriam S. Chaiken and Anne K. Fleuret. Westview Special Studies in Applied Anthropology, Westview Press Inc, Boulder, Colorado, USA and Oxford, 1990. I am most grateful to the editors and the publishers for permission to reprint the paper here. I am indebted to Tim Ingold and Alula Pankhurst for references and to Ato Dadafo Godana for translation of Defaye's book on Arussi.

Every family owned some cattle and sheep and goats, and an increasing number were acquiring herds of camels. In order to make best use of the sparse grazing, browse and water, a family usually divided its stock into distinct and territorially separated grazing units—one of milking cows, one of dry cows, one of milking camels and one of sheep and goats which run together. A very few exceptionally wealthy men had two or even three herds of milking cows, each in the care of a wife. According to the folk ideal, one of a set of brothers, or failing that close agnates, should be in charge of each unit according to seniority, that is the eldest with the milking cattle and the youngest with the sheep and goats. In practice, of course, this ideal was seldom achieved though almost invariably approximated; but we need not be concerned here with the ways in which the labor and stock of families were actually allocated (Baxter, 1966). For our present purposes it is sufficient to note certain consequences which followed from this particular style of animal husbandry. First, a family and its stock could well be distributed across an area as extensive as Wales. Secondly, the villages in which the milking cows were kept (*ola*) consisted mostly of older men, each with their wives and younger children; each elder was also the "father" (*aba*) or senior manager of the widely dispersed stock holdings of his family group.

A village of milking cattle, then, was also an assembly of elders, each of whom was, as it were, the center point of a set of specialized herding units; the members of any specialized unit were much more likely to visit the milking village than they were any of its satellite camps. The homestead of the milking herd was the place at which family plans were discussed and decisions made; the village of milking homesteads was the place where rituals were performed and the enduring moral point to which those who herded the dry stock in the satellite camps were attached. Milking villages and the dry stock may have been several days walk apart, but information and intelligence flowed readily between them. The members of a dry stock (*fora*) camp would be young childless couples, unmarried youths and girls and a more experienced man or two to advise. Camps moved frequently, especially following localized rainfall in the rainy season, whereas milking villages were much more settled;

indeed, in the favoured areas, such as the higher parts of Marsabit Mountain where the observations reported below were made, they were transhumant; villages moved down the Mountain to the plains in the rains and up again to the permanent wells for the dry season. If it did not rain in the plains, the milking villages remained where they were.

As cows matured they were brought up from the plains into the milking herds, where they normally remained, enjoying the richer grazing, until they ceased to bear calves. I was told that stock kept on Marsabit Mountain lost condition because there were no mineral springs (*hora*) for them to drink at, and because the richer mountain grazing could be too "strong". As a remedy, they were fed with *mulugdi* (Venturino, 1973, uses the spelling *muludde,* his ear is probably better than mine. I cannot locate the word in any other Oromo dictionary). *Mulugdi* are chunks of coarse salt and congealed sand collected from the surface crust of the Chalbi desert, which is a vast dried salt lake to the northwest of Marsabit (Dixey, 1948:3). The Chalbi becomes waterlogged after heavy rains, but, for most of the year, it is a baked, white, shimmering flatness, relieved only by mirages, and firm enough for lorries to drive across. When the father of a herd thought his charges needed *mulugdi,* he could send word to young members of his family grazing near the Chalbi to bring some up, or send down a couple of members of his village family with some donkeys and goatskin sacks to fetch some up. *Mulugdi* was freely available and collecting it was merely a rather tedious task. As I recall each homestead fetched its own supplies, though neighbours usually travelled together for the company. Negligible social planning or organized co-operation were required. Such small numbers of travellers would be welcome to spend the night at any camp or village, or could sleep safely and comfortably in the bush. Then the journey from Marsabit Mountain did not require passage through the territory of hostile or unfriendly peoples.

I feel confident that no Boran would have denied *mulugdi* to a neighbour in need, but I do not recall ever hearing anyone beg it, and it would seem out of character for a father of a herd to have done so. *Mulugdi* was on sale in Marsabit's embryonic little market for non-Boran townsmen to purchase (though I neglected to note the price). Before being served to stock it was broken up and moistened; I did see sheep fed it on a discarded sleeping hide, but it was usually served in a trough. All stock clearly relished *mulugdi* and would continue to lick the trough long after the last crumb had been eaten. *Mulugdi* was a valued and essential, but free, element in the animal husbandry of a stock-owning family. Boran from Marsabit District did not trade *mulugdi* nor natron (*magardo*), but Boran to the north did exchange it with Konso for grain, cloth and metal goods. In the last century Boran from the Mega area, in what is now southern Ethiopia, traded salt from Sogida (Salt) and natron to as far away as the Benadir coast; but this long distance trade had long ceased (Pankhurst, 1968:243, 436, 440; Abir, 1970:133). The famous *amole*, or salt bars, which were traditional articles of exchange, and were still occasionally to be seen in markets in southern Ethiopia in the 1960s, have no connection with either the salt trade or the collection and distribution of cattle-lick (Pankhurst, 1968:260–4).

I thought little about *mulugdi* then or later; it just seemed a common sense part of animal husbandry. Indeed I would probably not even have noted it, nor recalled it, but for the obvious pleasure people took in giving pleasure to their stock which, in turn, communicated itself to me. I had seen salt licks in the cattle fields and mineral blocks in the stalls of milking cows in England, and was vaguely aware that salt and mineral licks were widely used in stock management. I later learned that it was a common feature of animal husbandry across the world.

Ryder cites the description by Youatt (1840) of the stock management practices of the famous Spanish transhumant shepherds, which had "changed little over several hundred years ... on arrival at the summer pasture the *mayoral* (shepherd in charge) spread salt on stones for the sheep to lick to guard against possible ill effects from the sudden change to rich herbage" (Youatt in Ryder, 1983:432-3). Pratt and Gwynne (1976:162) note that East African sheep "make regular use of salt licks" and "it is generally agreed that cattle benefit from salt and mineral mixtures" (199). The *cure salee* was a regular part of the transhumant cycle in the Southern Sahara. In his masterly survey of salt production and trade in the central Sudan, Lovejoy summarizes the general animal husbandry practices as follows; "*Gari* [impure loose natron] was given to all livestock; donkeys, cattle and camels were fed the natron in their drinking water, while horses, sheep and goats were fed it with their food. *Gari* could be spread on the pastures for cattle. For horses, the *gari* was added to leaves ... Sometimes ash was added. The *gari* was mixed with millet or guinea corn chaff as a food for sheep and goats."Lovejoy cites Lander, Bowdich and others to show that the practice was general from Agades to Asante (1986:29). Messing notes that the Amhara of the Ethiopian plateau follow Oromo practice, "Fully grown steers are fattened by feeding them a mineral salt, for two years, or when mineral springs (Oromo: *ambo*) exist by taking them to water there (Oromo: *hora*, lake). This is done also to increase the milk supply of cows. 'Rejuvenators', consisting of large pieces of mineral salt, are fed to them for this purpose" (1985:50). It is worth noting that Messing's Amhara informants used Oromo in this context. Sato records of the Rendille, who are neighbours of the Boran and share a common habitat: "Salts are indispensable nourishment for all livestock, and the lack of regular salt may result in a low resistance to lameness, trypanosomiasis, as well as producing a general decline in activity. Thus the Rendille intentionally lead their livestock, especially camels, to waterholes with brackish water as well as to salt licks. When watering camels, the Rendille also keep them feeding on salt-shrubs such as *Dasysphaera prostata* and *Salsola dendroides africana*. Salts are also supplied to camels on the way to waterholes" (1980:17, 65).

I spent the academic year 1968–9 in the field among Arssi of Kofele District, Chilalo Province in Arussi Governorate in Ethiopia. The Arssi, like the Boran, are one of the several Oromo-speaking peoples who make up the Oromo nation; the two dialects are close and cultures similar. Arssi grew barley and kept stock. The eastern part of Kofele District and the western part of the neighbouring district of Gedeb formed a complementary economic area; the Kofele segment mostly lies between 2,485 and 2,970 meters and the Gedeb segment mostly between 2,000 and 2,485 meters (see Kebede, 1967). Generally speaking, cattle thrive better in Kofele during the dry season, because there is always water nearby, but better in Gedeb during the rains. Kofele produces cabbage, *ensete*[1] (false banana, *wuurki* in Oromo), bamboo for houses and stock pens, honey and other highland products; Gedeb produces more milk (and hence butter) and barley and a little flax as a cash crop. There are no mineral springs in either section. The Arssi were much more tied into the market economy than the Boran had been, and were much more restricted in their movements. All land was owned or tenanted and jealously guarded against intruders; most Arssi, in effect, were tenants of Arssi "big men" or Shoan landlords. But they still resided in cluster of agnates on land which they saw as lineage land, and all territory was still spoken of as "belonging" to particular specified lineages.[2] The day when homesteads could move freely with their herds, and any homestead had access to as much grazing as its herds could consume, was now only a folk memory. Pasture was in even shorter supply than stock and, like plow land, was guarded closely. But Arssi still asserted, as vehemently as ever Boran did, that they loved cattle and the dominant cultural ethos, as expressed in prayers, greetings, blessings, songs, and casual conversations, was a pastoral one. Most families from Kofele also maintained a homestead and land rights in eastern Gedeb, to which all or part of the family moved as good husbandry required. Arssi held very firmly indeed that cattle, small stock, horses, and mules, if they were to thrive, all needed regular access to mineral springs (*hora*) or to *boojji*, which is a salty mineral deposit similar to mulugdi, and which I shall describe more fully shortly.

I do not recall hearing the word *boojji* used in Kenya by Boran, but that may well be that I just did not note it. Certainly it was used, rather than mulugdi, in the market in Neghelli in Boran country in Ethiopia which I visited briefly in May 1969. (The price in the market there was 50 per cent higher than it was in Kofele.) Father Leus of the Roman Catholic Mission at Yabello, which is in Borana in Ethiopia, includes *boji* in his recent dictionary (1988) but does not give *mulugdi;* whereas Father Venturino, in his Kenya-based Borana dictionary (1973), gives *muludde* but not *boojji*. Both Arssi and Boran, when I asked, said, rather dismissively, that the two substances were "the same" or "resembled each other" and served the same purpose. I do not think the differences are material here. Certainly, all stock have a passion for *boojji/mulugdi* and both Boran and Arssi, as I suspect all Oromo, enjoy feeding it to their stock.

With his first pay the small boy my wife and I employed about our compound in Kofele went to the market, and spent ten cents on a heaped enamel bowl full of *boojji* for his beloved grandmother's few treasured sheep: my research assistant, Dadaafo Godaana, also bought *boojji* for his sheep, which were in the care of his father, and enjoyed taking me to see them enjoy it. When the late Adami Hussein's bedraggled horse, which was the apple of his eye, joined our compound, he gave it *boojji* to eat, which he had already bought in anticipation, even before grooming and drenching it. Horse and owner were clearly very happy.

I often saw *boojji* fed to stock in Kofele, where it was moistened and crumbled and evenly scattered on a clean patch of grass. The stock always licked and nibbled the grass and earth away until they left a shallow, bare depression. I was told that in areas which had a different type of soil, the *boojji* [3] needed to be mixed with a sort of red soil known as *hii'a*[4], in the proportions of one basket of *boojji* to two of red soil. Arssi asserted that they had "always" given *boojji* to their stock and that they had "always" gone to great efforts to ensure a regular supply of it. One indication of the importance in which it was held is that Tutschek[5], who compiled a dictionary of Oromo in the early 1840s in Munich, was given the word by his informants, who were freed slaves, homesick in an alien land, which they translated for him as "a kind of eatable earth". This suggests that, for the informants, the earth was an essential ingredient. The use of *boojji* is certainly not recent. Yilma Deressa, an Oromo historian, argues that one of the attractions of the Rift Valley lakes (and places such as Ambo) for Oromo settlers in the sixteenth century, was the presence of mineral springs and mineral licks *bowji* (1966–7:216).[6]

Here I must divert for a moment and comment on the word *hora*, "mineral waters", "source of mineral water", "hot spring": Gragg (1982) states that, in its primary meaning "hot spring water", the word has connections to both Arabic and Amharic. In

Oromo, it is a "complex word" in the Empsonian sense, with a wide field of meanings, from the casually mundane to the holy, which merits a paper to itself. Briefly, as a noun *horin* it may mean cattle, money, wealth, possessions; indeed all desirable and enduring material things. As a verb, in various forms, it may mean be rich, multiply, bear fruit, prosper, increase. *Hormata* are the reproductive parts of females and all other good things in life. In various forms it is used in blessings and as a response to prayers. It carries heavy charges of meaning. Water from *hora* is used in ritual purifications and, as Bartels puts it, "a *horra* means life in cattle and cattle is the life of the people" (1983:73). It is the *hora,* mineral water, that gives *boojji* its quality and its savour. But, as far as I could make out, *boojji* itself is simply a derivative and, in and as itself, does not have more than nutritional and medical properties. *Boojji* is just the name of a useful substance.

On dry Wednesdays there were usually a few *boojji* sellers at Kofele market—the most I noted was nine. On wet days there was no point in sellers coming, because *boojji* needs to be kept out of the rain so that its savour is not washed out. The going rates were 10 Ethiopian cents for a single heaped enamel bowl, or 50 Ethiopian cents for 20 heaped basins or flat basket lids plus two extra handfuls for goodwill; i.e. about three gallons for about 10 pence sterling or 20 American cents. In the market at Asasa town, the administrative center for Gedeb and further from the sources of *boojji,* the rate was 20 tins for 50 cents; to my eye the tins looked smaller than the basins used in Kofele. At Allellu Market, only a kilometer or so from Shashamene, and only half as far from the sources, the price was only 5 cents a bowl. These few figures suggests that the selling price increased the further away from the sources, as would be expected. But it was mostly only townspeople and the poor who bought *boojji*; usually in Kofele a group of neighbours, who would also be close agnates, would get together to make up a caravan of pack horses and pack mules to go to Lake Shaalla, Lake Abiata or Lake Langano to collect their own. Lake Shaalla was the nearest and most favoured source of *boojji* for people from Kofele. Caravans from Kofele were usually small, just three of four leaders and two or three times that number of pack animals.

Donkeys were rarely used, because they were slow and could only manage a half load. The driest months, March and April and September and October, were the times to travel, and then there was a flow of small local caravans and large caravans from Bale and Gedeb on the road. I lived alongside a major route and was in a good position to observe them. At trail junctions, such as the bridge over the River Webi, enterprising women set up charcoal braziers and sold boiled cabbage, barley cakes and coffee to the horse leaders and muleteers.

I visited two different *boojji* collecting points on Lake Shaalla Diika (Small Shaalla) on 2 May 1969, and one on Lake Shaalla Guddaa (Big Shaalla) on 24 May. The Small Shaalla collection points were on a goodish road, just below some hot springs and geysers which had become an attraction for tourists from Addis Ababa. *Shaalla* in Oromo means "pelican", with which the lake abounded and from which is took its name[7], and the area had been scheduled as a Game Sanctuary. The rocks and pools were littered with hulled maize cobs, chicken bones, and eggshells left by picnicking tourists who had tried cooking in the boiling pools. The steam which arose was believed to have curative properties, and Ethiopians came from great distances to sit on logs over the jets of steam. Though *boojji* permeated with the mineral water was valued so highly, stock were not brought to drink the waters. Though, I was told, they did go to drink at the cooler mineral springs at Lakes Zwai and Langano.

One of the labourers employed at the site, which was on the shoreline of a shallow bay, pointed out the shelters used to store *boojji*, but stated he was not expecting many, if any, more customers as the rains had already started two days earlier. My companions then told me that Little Shaalla was only a small site anyhow and that the big one at Big Shaalla was much busier! The *boojji* was sand and soil over which the mineral water from the hot springs had spread out and seeped in. *Boojji* had to be collected after a longish spell of dry weather. It was first raked and scraped into small mounds about two feet high to dry (cf. Lovejoy, 1986:61), then into larger mounds, which were covered with dom palm mats to protect them from the rain. Most was sold directly from the heaps but, as the rains approached, any left unsold was carried in baskets by labourers to the shelters.

Arssi stressed that *boojji* must always be protected from rain to prevent it being "destroyed", i.e. its savour and strength being washed out. In the market buyers tasted before they bought in order to check that the *boojji* had not lost its savour.

There were two collecting and selling points at Little Shaalla, both of which, I was told, were owned by absentee Amhara landlords. Amhara were the politically dominant ethnic group in Ethiopia, and Arssi expressed particular resentment at having to pay those whom they perceived as conquering and exploiting foreigners for that which God had given, as he had rain and grasses, to all men; or at any rate to all Oromo! The price was 50 Ethiopian cents a large gunny or hide sack. One sack made a mule or horse load. My very rough estimate is that one load would make about four 50 cent market lots of twenty basins each. The round trip from Kofele took four to seven days; so that the most that a man could make, for a week's exhausting work for himself and two pack animals, would have been just about 50 pence or one American dollar! The local rural economy was a desperately poor one.

I was not able to visit the Big Shaalla sites until 24 May, by which time the season was over. The approach road was still firm but looked as if it would be a quagmire in the rains, which made it clear to me that, even if the *boojji* could have been kept dry in the rains, the caravan paths and the streams and rivers they crossed would only have been negotiable with difficulty. By contrast, the paths to the Chalbi in Marsabit would have usually been open for 11 months of the year.

The organization of a small *boojji* collecting caravan from Kofele was fairly simple and straightforward for a homestead, or set of neighbours, which had sufficient men and/or child-free women fit enough to undertake the walk (or who had access to saddle horses) and the requisite pack animals. Many poor homesteads did not have pack animals, or only had one, so they had to borrow and, in return, lead a second or third animal on behalf of a wealthier homestead. This was not perceived as exploitation, because a homestead which had fewer horses was not likely to have a large herd or flock, so would not need so much *boojji*! This light patronage was wrapped in the egalitarian idiom of agnatic kinship; during the journey, as at home or when herding, members treated each other as kinsmen, deferring in speech and in behaviour only to age, generation and gender differences; no one was treated as an employee, servant or client. The parties were a little more formally structured than those of the Boran to the Chalbi for *mulugdi;* because all land was owned and guarded and because Kofele could be cold at nights, a party aimed to overnight at homesteads where they would be known and made welcome, so they needed friends or affines en route. But a small caravan was still a small group of kinsmen-neighbours. No one could order anyone else about.

This contrasted with the caravans sent for *boojji* by Amhara landlords, which were made up of their tenants and labourers. To me, as they passed along the road, they were indistinguishable from Arssi caravans, but to Arssi they were constituted quite differently, because the members of landlord caravans had not come together as kinsmen but as clients or servants. Foolishly, I never inquired where they overnighted.

There were only a few large Arssi landlords in Kofele; one did not exercise his rights as a landlord at all and behaved as nearly as possible like a traditional elder; another acted as much like an Amhara landlord as he could get away with; two others, who were outstanding innovators in many other ways, combined to send a lorry. But, in Gedeb and the neighbouring Bale Province, there were several wealthy stock-owners and landholders, spoken of as "big men", *nama guddaa,* who were descendants of those preconquest "big men" who had survived the conquest of the Arssi by the Abyssinian armies of Menilik. These men had also often acquired titles to large tracts of land, the status of government recognized head-men of a major clan section (*balabat*) and, quite often, an honorific Imperial title. These men, as their ancestors had done, organized large caravans, often of seventy or more pack animals; I saw one of well over a hundred animals. Counting was not easy, the caravans straggled and at bottlenecks, such as the Webi River bridge, intermingled. The men, and they were always men on the large, long distance caravans, were clansmen of the "big men" who had organized the caravans, and who might, or might not, accompany the caravan themselves. The clansmen would also be the tenants, at least in law, of the "big men" but, I was assured, most of these "big men" still acted much as their grandfathers had done; that is, as "fathers" (*aba*) or traditional leaders (*haiyu*). They did not collect rent for pasture in cash nor kind and neither paid nor ordered men to accompany them. (One young, schooled *balabat* was an exception to this and behaved, as he was legally entitled to do, as much like an Amhara landlord as he dared.) I was told, on one hand, that one mark of a "big man" was that he could organize large caravans and, on the other, that one needed to be a "big man" in order to have the resources to organize the caravan. But, certainly, not all "big men" organized large caravans; to do so one needed junior clansmen to participate. After the conquest clansmen had been redesignated tenants, but prior to that they would have been free equals; some "big men" still behaved according to the preconquest ideal, others according to the colonial ideal of the armed settler (*neftanya*) landlord. I was not able to get data on the detailed social composition of any large caravan because they originated forty or more miles away from where I was living, and merely passed through Kofele.

But the point I want to make here seems clear enough: "big men" needed to be able to organize the labour of a number of poorer men, as well as to mobilize the homesteads of other wealthy clansmen, because a caravan needed pack animals and leaders. To travel together in a caravan made for sociability and very welcome security. Arssi had never participated in an encompassing national "Peace" (*Nagaa*), such as Boran enjoyed (see Baxter, 1965), and Gedeb and Bale was not safe nor secure, so a large caravan supplied the safety that lay in numbers and the protection of a "big man's" name. But a long distance caravan had logistic as well as security problems; travellers could not rely for hospitality on friends or affines en route, and their numbers would have been more than most compounds could entertain. A "big man" had to rely on similar "big men" at appropriate stages along the caravan route, so that his men and pack animals could be fed, watered and be kept safe overnight. Arssi "big men" formed a loosely connected network across Arssi country, and even into the fringes of the national ruling elite, linked by affinity through the exchange marriage of sisters and by matrilineal kin ties. The most conspicuous attribute of "big men" was the number and span of their many marriages; indeed many could not be constrained to the Islamic maximum of four. In this respect "big men" were quite unlike even very wealthy ordinary men. To get a large, long distance caravan on the road, then, took stock-wealth, widespread and effective affinal connections, and organizational skills; organizing a caravan did not make one a "big man", one had to be a "big man" first in order to do it.

What Arssi stressed as the attributes of a "big man" were not so much his wealth, nor his wives, nor his power, but the social tact or knowledge of words which he required in order to get so many vociferously independent homesteads to co-operate. Power and wealth, in themselves, were not sufficient to enable a "big man" to organize a caravan, simply because blatantly exercised power, indeed even the slightest appearance of being overbearing, would have dispersed the latent social power that wealth and the other attributes allowed.

The names of former "big men", of course, are remembered because they have had descendants who have become "big men"; but what was always recalled about the "big men" of old was their abilities as organizers of long distance caravans to Welayita and Wollamu for cloth, to Bale and beyond for salt[8], and to the Rift Valley Lakes for *boojji* (see Baxter, 1984:462). They were not remembered as cloth *traders*, not as salt *traders*, let alone as *boojji traders*; they were organizers not merchants. They are not honoured in memory for their wealth nor their power but for their organizational skills, which they used, like they did their skill with "words", for the general benefit.[9] I am not suggesting that nineteenth century and turn of the century "big men" were just simple, public-spirited benefactors, but their organizational capabilities were represented to me, correctly I think, not as stemming from coercive power so much as from the abilities to manage men and to manage words.

This is not to say that power and wealth were not prerequisites; certainly no poor man could have made a name as a caravan organizer as he could, for example, as a warrior, but just that, in themselves, power and wealth were not sufficient; nor was the pursuit of wealth an end in itself. It is reasonable to assume that many wealthy and powerful men in the past, as in the present, were not able to organize their fellows at all, and thus become remembered as "big men"; indeed, they are only recalled at all in the genealogies of their immediate descendants. One could not be a "big man", as it were, in a little pool. In eastern Gedeb and in Bale in 1969, "big men" still had a role as organizers of *boojji* caravans, even though their roles as organizers of cloth and salt caravans had been usurped by traders with lorries. In Kofele, which was densely settled, secure and close to the sources of *boojji*, "big men" as organizers had become redundant.

As *boojji* is seen as an essential component of animal husbandry, and hence of Arssi identity as dedicated stock herders; so, the "big men" who organized its collection were represented to me as essential components of Arssi historical identity. But status cannot be adapted to other times and situations. The descendants of "big men" in Kofele, however wealthy and honoured by the Ethiopian administration, could not be made into contemporary community leaders nor be respected as "big men", simply because now they contributed nothing to the common weal. Whatever their coercive and economic powers, they were simply wealthy landlords and government headmen *balabats*.[10]

Finally, a postscript on indigenous knowledge or, perhaps, an epitaph for *boojji*. It must be clear that Arssi and Boran and, I think, most Oromo and, indeed, most inhabitants of Southern Ethiopia, hold that *boojji* is essential to the well-being of stock that do not have access to mineral springs. Certainly non-Arssi in Kofele shared this belief, and I have mentioned that Amhara landlords organized caravans to collect *boojji* and that non-Arssi townsmen bought it in the markets. The official handbook of the Arussi Governorate, which was compiled by Amharas and Shoans, lists its mineral springs and sources of *boojji* among the natural resources of the Governorate (Defaye, 1968:157).

I assumed that the Arssi and others knew what they were about and did not put themselves to great trouble for nothing. Most stockmen across the world have also thought salt and mineral licks good for their stock. As the *African Veterinary Handbook* put it: "it is essential to provide cattle kept on limited grazing with salt and mineral licks" (Mackenzie and Simpson 1953, 1971:214). Pratt and Gwynne have since put it very succinctly: "In most parts of East Africa livestock have a craving for salt, and there are numerous natural licks which have been used by wild and domestic animals for centuries. Few of these natural licks show the advantage when analyzed, and it has often been noticed that they are abandoned when animals have access to an introduced mineral mixture, well suited to the area" (1977). They also point out that licks are "a useful tool in grazing management" (Pratt and Gwynne, 1977:199). This experienced pair of rangeland scientists also note that "most pastoralists are good stockmen", and that, "in certain localities one or the other food salts may be deficient. A deficiency which will reveal itself in low fertility of breeding stock, weakness of the calves and slow growth of the young stock" (*ibid.*, 1977:152). These remarks might well have been made by Boran or Arssi.

The importance of *boojji* to the welfare of the stock seemed obvious to me, but I was quite incapable of even guessing in what its importance lay. I asked the Chilalo Agriculture Development Unit[11], a sensitive Swedish-Ethiopian development organization that was working in the area, if they knew what *boojji*` actually contributed to the wellfare of the stock. They were, as always, most helpful; they had already had *boojji* analyzed and had concluded that it made no physical difference at all to the stock who could manage quite well without it. This view was not, of course, taken by the stockmen employed by the Development Unit and it was they, the workers at the very bottom of the hierarchy, on whom the stock depended for their day to day care. The stockmen were the only actually indispensable workers on the project. Like the old Buryat shepherds in the Siberian Karl Marx Collective described by Caroline Humphrey (1983:231), their traditional knowledge ("golden words") and skills were admired, but also devalued because their carriers were unschooled.

I found this "scientific" dismissal of *boojji* hard to accept. The Oromo, like any people anywhere at any time, I felt, might have been misled for centuries by the supposed mystical attributes of a medicine or a food, but *boojji* did not have any mystical qualities attributed to it. Neither its collection nor the way it was fed had any mystical overtones. Both Boran and Arssi were sharp-eyed, open-minded, experimentally oriented stockmen; if they asserted that stock deprived of water from mineral springs or *boojji* lost condition, I felt certain that they were likely to be right. In an inarticulate sort of way I felt that respect for indigenous knowledge was not just, "a courtesy to the people ... [but] an essential first step to successful development" (Brokensha *et al.*, 1980:8), and should be heeded. So I took the question on to my friend Dr. Hugh Harding of the Veterinary Training School at Bishoftu. He suggested that *boojji* probably contained trace elements which could not then be discerned by laboratories in Ethiopia. He offered to take samples to FAO for analysis. I sent him specimens just as he was leaving so they would be fresh. But we then learned that the samples were classified as prospectors' mineral samples and hence would need an export certificate, which there was not time to obtain. When I came to leave I did not attempt to get permission to export samples, for a variety of reasons.

As far as I can ascertain *boojji* has still not been analyzed. I still find it hard to accept that, perhaps like some academic papers, it is merely a dummy, a comforter, a useless formula, a placebo, or just plain wrong. At least, so far at any rate, no one has suggested that *boojji* is harmful. But, of course, "truths" change, especially those concerned with diet and nutrition. I was brought up to believe as revealed truth what is now held to be self-evidently false; i.e. that a healthy human diet has as its basis full cream milk, farm-churned butter, home-cured fat bacon and ham, eggs, dark, rich dripping toast and suet dumplings that stuck to the ribs, and that white bread was purer than wholemeal. Similarly, cattle thrived on clean bedding, dry standings, sweet hay, deep pastures, and sweet air!

But, of course, the scientific status of *boojji* is entirely irrelevant for us as anthropologists, as are the scientific statuses of homeopathic medicine, the cures of spiritual healers, or the biological consequences of the rules which define incestuous relationships or food prohibitions. Shared knowledge of such sorts is a community construction which is quite unconcerned with measurable "truth". The heartfelt, belly-warming responses (*garaa wollja laachuu*) generated between human providers of *boojji* and its animal receivers have a value which is independent of the presence or absence of trace minerals. The consumption of *boojji* was an occasion for shared animal and human conviviality and joy. It is a cyclically repeated representation of the mutual dependencies of the stock of a homestead and the people of a homestead (*warra*). I have already noted the complex resonances of *horin* "productive (i.e. reproductive) stock, wealth"; *warra* "people, family" is an equally complex and resonant word. Here, the simple point is that, in proper times, you cannot have one without the other. Stockless individuals, whatever their genealogical connections, are not a *warra*. Stock join people. That I should have been concerned about the nutritional value of *boojji* only demonstrates my own ethnocentric, intellectual myopia, which led me to concentrate on the object, rather than the relationships which it generated.

Across the world those pastoral economies and cultures which have not already been destroyed are threatened with extinction and the arid zones with consequent degradation. One contributory reason for this ecological and human catastrophe is just that the powerful everywhere perceive livestock simply as a commodity, dull and inert like battery hens, mineral ore, stocks and shares or figures in a balance sheet. Whereas, in truth livestock are active generators of social relationships, exemplars of truth and beauty and stimulators of the poetic imagination, of living memory and of the informed heart. Indigenous knowledge and practice, any more than "ethnographic facts", are not isolable snippets of useful knowledge which can be pulled up, or out, and just added to the world's store of intellectual bric-a-brac.

NOTES

1 Convenient descriptions of *ensete* cultivation etc., are given in Smeds (1955) and Shack (1966), Chapter 3.

2 For thumbnail sketches of the political and economic situation in Kofele District at the time of the fieldwork, see Baxter (1980, 1983, 1984). The situations have, of course, radically changed since the Revolution.

3 I follow the spelling of Gragg's Oromo dictionary (1982) here. Gragg translates *boojji* as "salty sediment left after water evaporated (cows like to lick it)". Tutschek writes *bodji* which approximates more to the Arssi pronunciation than does *boojji,* he also gives *handjiro* and *goe* as sorts of "eatable earth", both of which looked something like grain. Leus (1988), a priest who is working in Boran, writes *boji,* "salty crust on the ground ... cattle, kudu etc., eat and lick it".

I have wondered if it might be *boojji* that is meant by *bogito,* rather than "sprouts", in the following song collected by Cerulli (1922:148):

"1. abba farda collie	1. Master of the swift courser,
2. gababa botollie	2. short, little,
3. bogito nadita	3. thou eatest sprouts (bogito).
4. bogite galta	4. Thou wilt return a prisoner!
5. sombo gudda gala bai	5. Come forth under the great sycamore!
6. bode dina dura bai	6. Come forth before the lances of the enemies!
7. mie dawaccisi	7. Come, gallop!

This is a song of contempt of the shepherds for the horsemen who gallop passing near the flocks. *Bogito* (v. 3) is a green vegetable similar to the sprouts, also a food for the poor."

I cannot find the word *bogito* translated as "sprouts" in any dictionary or any other source, and I cannot think of any reason why shepherds should hold horsemen in contempt, or vice versa. Oromo are famed horsemen; in Jimma, for example, kings were named after their horses, and all Oromo are shepherds it is not a specialized occupation. It seems more credible that the horses should eat *boojji* to give them strength than eat "sprouts"; this is even more likely if the next line is translated as "bring back prisoners of war", which it could well be. The last three lines also sound more like praise than contempt. The punning and sound parallelism are typical of Oromo poetry. Further, one has to travel in a party to collect both *boojji* and *boojia* "captives, booty".

4 This sounds like Leus' *Haya-i,* "a better quality, more salty (*boji*): good for the cattle".

5 See Pankhurst (1976) for an account of the Oromo young men who reached Munich via Cairo slave market, and the conditions under which they assisted the estimable Tutschek.

6 I have not got a copy of Yilma Deressa's work to hand, and rely here on notes that I made in the Institute of Ethiopian Studies in Addis Ababa.

7 Gragg gives *haad'a waaqoo* for pelican, literally "mother of god". I did not hear this nor have I located it in any other Oromo dictionary; it may be restricted to Leka Oromo. The word sounds as if it may be semi-serious, semi-jocular, rather like *gaala Waakat* "God's camel" for the praying mantis.

8 Braukamper (1980:253) notes the importance of *boojji* caravans in this area. McClellan (1986:181) notes that neighbouring Gedo "could double the value of *ensete* by transporting it to neighbouring Guji [also Oromo pastoralists] and exchanging it for butter, cotton, livestock or *bole* (an earth salt fed to cattle)". Since I first wrote this paper I have been privileged to read a manuscript by Rev. van de Loo on the Guji Oromo which contains a wealth of ethnography. He notes that *duba* is a mixture of salt and sand given to cattle. "When the grass is growing again a mixture of salt, red lime, and water is given to the cattle in the long trough, *bidurru*." Guji organize caravans to undertake the dangerous and tiring journey to collect salt from Lake Abaya. van de Loo cites songs in honour of the journey. Guji have a *halusisisa* ritual for the horses before the start of the journey. In one song a lover was rejected because he failed to bring salt for the cattle. One text notes that a woman who fails to prepare good food for her menfolk to take on the journey "risks a whipping". As in the Sudan, salt for stock as for humans was one item in a complex of trade goods and was not a specialized trade. Few, if any, African traders got rich from salt trading alone.

9 Interestingly, Abdul Sheriff (1987:169) cites an unpublished Ph.D. thesis by Jackson (1972) to argue that Kamba ivory traders in the nineteenth century "constituted a 'small economic class' of powerful or prominent men. They did accumulate wealth to some extent in cattle, but Jackson was struck by the lack of emphasis placed on personal followings and clients".

10 Cf. Donham's account of Maale *balabats* who took on the lifestyle of their Amhara masters (Ch. 2, 1985).

11 See Nekby (1971) for a summary account of CADU and its activities.

REFERENCES

Abir M., 1970: "Southern Ethiopia", In *Pre-Colonial Trade,* D. Birmingham and R. Gray, eds, pp. 119–137, Oxford, Oxford University Press, *op. cit.*

Bartels Lambert, 1983: *Oromo Religion: Myths and Rites of the Western Oromo of Ethiopia. An Attempt to Understand,* Berlin, Collectanea Instituti Anthropos, Dietrich Reimer.

Baxter P.T.W., 1965: "Repetition in Certain Boran Ceremonies", In *African Systems of Thought,* M. Fortthe es and G. Dieterlin, (eds), pp. 64–78, London, Oxford University Press for International African Institute.

Baxter P.T.W., 1966: "Stock Management and the Diffusion of Property Rights". *Proceedings of the Third International Conference of Ethiopian Studies,* pp. 116–127, Addis Ababa, the Institute of Ethiopian Studies, Haile Selassie I. University.

Baxter P.T.W., 1980: "Always on the Outside Looking In: A View of the 1969 Ethiopian Elections from a Rural Constituency", *Ethnos,* 1–2:39–59.

Baxter P.T.W., 1983: "The Problem of the Oromo or the Problem for the Oromo". In *Nationalism and Self-Determination in the Horn of Africa.* I.M. Lewis, (ed), pp. 129–149, London, Ithaca Press.

Baxter P.T.W.,1984: "Butter for Barley and Barley for Cash: Petty Transactions and Small Transformations in the Arssi Market". In *Proceedings of the Seventh International Conference of Ethiopian Studies,* Sven Rubenson, ed, pp. 459–472. Uppsala, Sweden and East Lansing, MI, the Scandinavian Institute of African Studies and the African Studies Center, Michigan State University.

Braukamper Ulrich, 1980: *Geschichte der Hadiya Süd-Äthiopiens: von den Anfangen bis zur Revolution* 1974, Wiesbaden, West Germany, Franz Steiner.

Brokensha David, D.M. Warren, and Oswalt Werner, (eds), 1980: *Indigenous Knowledge Systems and Development.* Lanham, MD, University Press of America.

Cerulli Enrico, 1922: *Folk Literature of the Galla of Southern Ethiopia,* Vol. III, Cambridge, MA, Harvard African Studies.

Defaye Dejazmatch Sailu, 1968: *Arussi Population, Development and Community,* Addis Ababa, Assela.

Deressa Yilma, 1966–7: *Imperial Ethiopian History in the Sixteenth Century,* Typescript. Addis Ababa.

Dixey F., 1948: *Geology of Northern Kenya,* Geological Survey of Kenya No. 15, Nairobi. Government Printers.

Donham Donald, 1985: *Work and Power in Maale, Ethiopia,* Ann Arbor, MI, University Microfilms Inc. Research Press.

Gragg Gene, 1982: *Oromo Dictionary,* Monograph No. 12, East Lansing, African Studies Center, Michigan State University.

Humphrey Caroline, 1983: *Karl Marx Collective: Economy, Society and Religion in a Siberian Collective Farm*, Cambridge, Cambridge University Press.

Jackson K.A., 1972: *An Ethno-Historical Study of the Oral Traditions of the Akamba of Kenya*, Unpublished PhD Dissertation, University of California, Los Angeles.

Kebede Yilma, 1967: "Chilalo Awraja", *Ethiopian Geographical Journal*, 5(1):25–36.

Leus Ton, 1988: *Borana–English Dictionary*, typescript, Dadim, Ethiopia, Dadim Catholic Church, Ethiopia.

Lovejoy Paul E., 1986: *Salt of the Desert Sun: A History of Salt Production and Trade in the Central Sudan*, Cambridge, Cambridge University Press.

McClennan Charles W., 1986: "Coffee in Centre–Periphery Relations: Gedo in the Early Twentieth Century", In *Southern Marches of Ethiopia*, D. Donham and W. James, eds, pp. 175–195, Cambridge, Cambridge University Press.

Mackenzie P.Z. and Simpson, R.M. 1953 and 1971: *The African Veterinary Handbook*, London, Pitman.

Messing Simon D., 1985: *Highland Plateau Amhara of Ethiopia*, New Haven, Conn., Human Relations Area Files Flex Books.

Nekby Bengt, 1971: *CADU: An Ethiopian Experiment in Developing Peasant Farming*, Stockholm, Prisma.

Pankhurst Richard, 1968: *Economic History of Ethiopia 1800–1935*, Addis Ababa, Haile Selassie University Press.

Pankhurst Richard, 1976: *The Beginnings of Oromo Studies in Europe*, Roma, Africa, XXXI(2):171–206.

Pratt D.J. and Gwynne, M. D., 1976: *Rangeland Management and Ecology in East Africa*, London, Hodder and Stoughton.

Ryder M.L., 1983: *Sheep and Man*, London, Duckworth.

Sato Shun, 1980: "Pastoral Movements and the Subsistence Unit of the Rendille of Northern Kenya: With Special Reference to Camel Ecology", Osaka, *Senri Ethnological Studies*, 6:1–78.

Schack William A., 1966: *The Gurage: A people of the Ensete Culture*, London, Oxford University Press for the International African Institute.

Sheriff Abdul, 1987: *Slaves, Spices and Ivory in Zanzibar: Integration of an East African Commercial Empire into the World Economy, 1170–1873*, London and Dar es Salaam, Tanzania, James Curry and Tanzania Publishing House.

Smeds E., 1955: "The Ensete Planting Cultures of Eastern Sidamo, Ethiopia", *Acta Geographica*, XIII(4):1–39.

Tutschek Charles: 1844 and 1845 Dictionary of the Galla Language, Munich.

Venturino Bartolomeo, 1973: Dizionario Borana–Italiano, Bologna, Italy, Editrica Missionara Italiana.

List of Contributors

M.E. Abu Sin
Geographer, Department of Geography,
University of Khartoum,
P O Box 321, Khartoum, Sudan

Michael Angstreich
Development Officer, CARE International,
P O Box 180, Sentrum, N-0102 Oslo, Norway

Odd Eirik Arnesen
Geographer, Department of Geography,
University of Trondheim,
N-7055 Dragvoll, Norway

P.T.W. Baxter
Social Anthropologist, Department of Social Anthropology,
University of Manchester,
Manchester, M13 9PL, Great Britain

Jerome O. Gefu
Rural Sociologist, National Animal Production Research Institute,
Ahmadu Bello University,
PMB 1096, Zaria, Nigeria

Mamman Aminu Ibrahim
Veterinary Scientist, Department of Physiology and Pharmacology,
Faculty of Veterinary Medicine,
Ahmadu Bello University,
Zaria, Nigeria

Stig Johansson
Forester, Department of Silviculture, University of Helsinki,
Unioninkatu 40B, SF-00170 Helsinki, Finland

George A. Keya
KARI, National Arid Land Research Centre,
P O Box 147, Marsabit, Kenya

M.A. Mohamed Salih
Social Anthropologist, the Scandinavian Institute
of African Studies,
P O Box 1703, S-751 47 Uppsala, Sweden

Mohamed Said Samantar
Economist, the Institute of Development Studies,
University of Sussex,
Brighton BN1 9RE, Great Britain

Camilla Toulmin
Agricultural Economist, International Institute for Environment
and Development,
3 Endsleigh St, London WC1 HODD, Great Britain

Seminar Proceedings
from the Scandinavian Institute of African Studies

1. *Soviet Bloc, China and Africa*. Eds. Sven Hamrell and C.G. Widstrand. 173 pp. Uppsala 1964. (Out-of-print)
2. *Development and Adult Education in Africa*. Ed. C.G. Widstrand. 97 pp. Uppsala 1965. (Out-of-print)
3. *Refugee Problems in Africa*. Ed. Sven Hamrell. 123 pp. Uppsala 1967. SEK 30,-
4. *The Writer in Modern Africa*. Ed. Per Wästberg. 123 pp. Uppsala 1968. SEK 30,-
5. *African Boundary Problems*. Ed. C.G. Widstrand. 202 pp. Uppsala 1969. SEK 30,-
6. *Cooperatives and Rural Development in East Africa*. Ed. C.G. Widstrand. 271 pp. Uppsala 1970. (Out-of-print)
7. *Reporting Africa*. Ed. Olav Stokke. 223 pp. Uppsala 1971. SEK 30,-
8. *African Cooperatives and Efficiency*. Ed. C.G. Widstrand. 239 pp. Uppsala 1972. SEK 60,-
9. *Land-locked Countries of Africa*. Ed. Zdenek Cervenka. 368 pp. Uppsala 1973. SEK 80,-
10. *Multinational Firms in Africa*. Ed. C.G. Widstrand. With an introduction by Samir Amin. 425 pp. Uppsala 1975. (Out-of-print)
11. *African Refugees and the Law*. Eds. Göran Melander and Peter Nobel. 98 pp. Uppsala 1978. SEK 50,-
12. *Problems of Socialist Orientation in Africa*. Ed. Mai Palmberg. 243 pp. Uppsala 1978 (Out-of-print)
13. *Canada, Scandinavia and Southern Africa*. Eds. D. Anglin, T. Shaw and C.G. Widstrand. 190 pp. Uppsala 1978. SEK 70,-
14. *South-South Relations in a Changing World Order*. Ed. Jerker Carlsson. 166 pp. Uppsala 1982. SEK 90,-
15. *Recession in Africa*. Ed. Jerker Carlsson. 203 pp. Uppsala 1983. SEK 95,-
16. *Land Management and Survival*. Ed. Anders Hjort. 148 pp. Uppsala 1985. SEK 100,-
17. *Religion, Development and African Identity*. Ed. Kirsten Holst Petersen. 164 pp. Uppsala 1987. SEK 110,-
18. *The IMF and the World Bank in Africa: Conditionality, Impact and Alternatives*. Ed. Kjell J. Havnevik. 179 pp. Uppsala 1987. SEK 110,-
19. *Refugees and Development in Africa*. Ed. Peter Nobel. 120 pp. Uppsala 1987. SEK 110,-

20. *Criticism and Ideology. Second African Writers' Conference—Stockholm 1986*. Ed. Kirsten Holst Petersen. 221 pp. Uppsala 1988. SEK 150,-
21. *Cooperatives Revisited*. Ed. Hans Hedlund. 223 pp. Uppsala 1988. SEK 170,-
22. *Regional Cooperation in Southern Africa. A Post–Apartheid Perspective*. Eds. Bertil Odén and Haroub Othman. 243 pp. Uppsala 1989. SEK 170,-
23. *Small Town Africa. Studies in Rural–Urban Interaction*. Ed. Johathan Baker. 268 pp. Uppsala 1990. SEK 170,-
24. *Religion and Politics in Southern Africa*. Eds. Carl Fredrik Hallencreutz and Mai Palmberg. 219 pp. Uppsala 1991. SEK 170.-
25. *When the Grass is Gone. Development Intervention in African Arid Lands*. Ed. P.T.W. Baxter. 215 pp. Uppsala 1991. SEK 170,-